The Architecture of Innovation

The Architecture of Innovation

The Economics of Creative Organizations

JOSH LERNER

OXFORD

UNIVERSITY PRESS

Great Clarendon Street, Oxford, OX2 6DP,
United Kingdom

Oxford University Press is a department of the University of Oxford.
It furthers the University's objective of excellence in research, scholarship,
and education by publishing worldwide. Oxford is a registered trade mark of
Oxford University Press in the UK and in certain other countries

First Edition published in 2012

Impression: 1

British Library Cataloguing in Publication Data

Data available

Library of Congress Cataloguing in Publication Data

Data available

ISBN 978–0–19–963989–2

Printed in Great Britain on acid-free paper
by Clays Ltd, St Ives plc

To Brodie

Contents

Acknowledgments

There are many important people to acknowledge. I owe a debt of gratitude to several research associates who helped with the development of this book: Lawren Tilney, Quan Tran, Abishai Vase, Ethan Waxman, and especially Andrew Speen. Chris Allen of Baker Library provided great help, as usual, with the researching of various facts and figures. My assistant, Maurie SuDock, helped the effort in innumerable ways. I also should acknowledge Harvard Business School's Division of Research, which provided financial support for the project.

At Harvard Business Review Press, my editor, Tim Sullivan, provided ongoing support and encouragement, as did David Musson of Oxford University Press. My agent, Scott Moyers of the Wylie Group (until his departure for the dark side of publishing), played a key role in conceptualizing and encouraging the writing of the book, as well as Adam Eaglin at Wylie. I likewise thank copyeditor Sarah Weaver and production editor Jen Waring.

Three other sources for this work should also be acknowledged. Many of the insights in the final section of the book, dealing with the vagaries of creating a hybrid between the venture capital and corporate research models, originally emerged from projects for firms. While it might be ideal if the knowledge in this space was completely codified, and one could refer to definitively documented academic studies, this is unfortunately not the case. The chance to work on these projects, and to understand the many barriers to implementation, definitely enriched my understanding of these issues.

This book draws on the ideas (and occasionally the words) in my previously published research, including work with Pierre Azoulay, Shai Bernstein, Paul Gompers, Felda Hardymon, Steve Kaplan, Sam Kortum, Ann Leamon, Tom Nicholas, and Antoinette Schoar. I thank them for permission to use here some of our jointly developed ideas, as well as for many conversations that shaped my own ideas. Pierre deserves particular thanks for carefully reading the manuscript and making many helpful suggestions. I also had helpful conversations with several friends and colleagues, including Ed Colloton, Ralph Lerner, Gustavo Manso, Ramana Nanda, and Scott Stern.

Finally, over the past decade, I have had the chance to organize a series of groups and conferences at the National Bureau of Economic Research, devoted to issues of entrepreneurship, innovation, and the settings where these activities can most successfully thrive. The NBER's chief executive officers, Marty Feldstein and Jim Poterba, encouraged these activities. Carl Schramm, Bob Litan, and Bob Strom of the Ewing Marion Kauffman Foundation were generous in supporting these groups. Those settings provided occasions for conversations with friends and colleagues about many of the ideas discussed in this book.

—*Boston, Massachusetts*
  April 2012

1

The Search for
Innovation and Growth

ET'S START WITH the stories of two acquisitions.

In August 2011, Google announced the acquisition of the bulk of Motorola.[1] No one had thought of Motorola as being particularly innovative in a long time, despite the firm's best attempts. Instead, for many years, the company had focused on filing successful patents. Employees who successfully filed patents were rewarded with a bonus of several thousand dollars. Researchers who made at least ten patent filings sported gold-colored employee badges, while those with at least twenty-five wore platinum-hued ones. Additional financial bonuses were associated with reaching these plateaus.

As a consequence, there was a strong emphasis at Motorola on making a large number of incremental patent filings. For instance, Motorola filed fifty patents in the late 1990s for battery latches on wireless phones alone, a feature that drives few consumers to decide whether to purchase a new phone. Each domestic patent filing typically costs tens of thousands of dollars, and international filings many times that amount. Battery latch patents were clearly a losing proposition, and yet still, there they were—most likely accompanied by some snazzy gold- and platinum-colored badges.

But a larger consequence of these policies—and Motorola's approach to innovation more generally—may have been even more problematic. The company failed to pursue the innovations that led to a fundamental change in its key markets in the ensuing years. After introducing the stylish superthin Razr mobile handsets in 2004, a huge success, the firm struggled to find a follow-up. In its core handset market, the firm failed to understand that consumer demand was shifting from basic phones with stylish design to more complex devices that could handle a wide range of functions, from getting directions to updating one's status on Facebook.

From 2006 to 2010, Motorola fell from number one to number eight in U.S. shipments of mobile handsets.[2] More generally, starting in the late 1990s, the firm made a series of costly blunders that seemed closely linked to a tendency to focus on certain technological avenues and not anticipate the changing nature of market demand: for example, its $5 billion Iridium satellite phone project, which ended in bankruptcy within a year of coming on line, and its competitive reversals in China, where it failed to anticipate the transition to digital technology and the GSM standard.[3]

So it was not surprising that when Google announced its acquisition, analysts argued that the major motivation for the deal was the search giant's desire to get its hands on Motorola's patent portfolio. The key attraction, rather than being any given product, was the large number of patents, which could be employed as a bargaining chip when negotiating cross-license agreements.[4]

The second acquisition started with a massive failure, at least as far as we traditionally measure business success: the peer-to-peer network Kazaa.[5] In 2000, Niklas Zennström and Janus Friis, along with a team of crack Estonian programmers, had developed the file-sharing service. Kazaa allowed users to exchange MP3 music files and other files, such as videos, over the Internet. Kazaa proved to be enormously successful, with over 300 million copies downloaded by 2003. Unlike centralized services such as Napster, by distributing much of the activity to its network users, Kazaa could grow quickly and inexpensively. In that year, *Fortune* reported, "Kazaa" was the top Internet search term, exceeding Harry Potter, the rapper 50 Cent, and even Paris Hilton.

Alas, all was not rosy. The recording industry soon noticed that users, rather than exchanging home recordings and videos of their pets doing tricks, were instead using Kazaa to busily trade Top 40 songs, movies, and other material that did not belong to them. Despite Zennström and Friis's protestations of ignorance, Kazaa soon became enmeshed in a web of litigation with the Recording Industry Association of America and other powerful adversaries. The founders, who had tried to distance themselves from their creation through a series of byzantine legal maneuvers, were soon on the run, seeking to elude the music industry henchmen who sought to serve them with subpoenas.

It was against this backdrop that Zennström and Friis came up with their next idea, in 2002: a peer-to-peer phone network. Initially, the two had planned to exploit devices that could communicate over Wi-Fi networks. But at the time, these networks—while today seemingly ubiquitous at every coffee shop and hotel on the planet—were still in their infancy. The two met with the individual who would be their initial venture backer, Howard Hartenbaum from Draper Investment (a firm founded by Bill Draper, the legendary founder of pioneering firm Sutter Hill), and refined their vision. Over the summer of 2002, the three men shifted to a focus on allowing calls from one personal computer to another, during which registered users could talk to each other for virtually nothing. Together, they realized they could work again with the Estonian programming team to build a network that would allow Skype, their new company, to grow rapidly without adding significant infrastructure, unlike traditional centralized Internet-based phone providers such as Vonage. Along with other early investors, such as Bessemer Venture Partners, Index Ventures, and Draper Fisher Jurvetson (run by Bill's son, Tim), Draper invested in and nurtured the firm. The rough idea was transformed into a product juggernaut. In 2005, as the firm's users hit seventy-five million, Skype was purchased by eBay for at least $2.6 billion.

The lesson from these two stories might seem perfectly clear: act more like Skype and less like Motorola.

If only it were that easy.

The Architecture of Innovation examines the critically important area of how we can spur innovation. Despite the importance of technological discovery, the process by which firms and nations innovate is shrouded in mystery. In many cases, misconceptions and misleading stories mask what is really taking place, ambiguities that can lead readers to conclude that innovation is a crapshoot, a random process in which a few get lucky but many fail. Is there any rhyme or reason why some firms, individuals, or nations are particularly innovative while others are not? If the answer is no, the possibility that we will address our critical economic and social needs may be slim indeed.

Innovation *can* be understood and managed. But many of the most useful insights have been largely neglected by the voluminous business literature on innovation and by many of those who seek to encourage innovation, whether top executives, investors, or policy makers. In particular, far too little attention has been paid to the power of incentives in shaping the behavior of those who design and commercialize innovations.

Making this gap particularly troubling is the fact that over the past two decades, economists have developed a rich array of insights into how incentives can work—and not backfire—in promoting new ideas. But many of these ideas have not yet crossed over into practice. This resistance may be a reflection of the backgrounds of a key audience: many of the people responsible for managing, investing in, and shaping innovation are scientists and engineers. In any case, the gap between the recent economic insights and practice remains dismayingly large. *The Architecture of Innovation* aims to change that.

Careful study into the ways organizations effectively innovate suggests the power of appropriate rewards. Without rewards, far too often good ideas languish unused, and researchers end up pursuing blind alleys. But creating incentives for innovation is not simply a matter of "more is better." Just as an architect can ruin the design of a building by adding too much ornamentation of different styles, too-powerful and misdirected incentives can introduce distortions that actually discourage new discoveries. As social psychologists and economists alike have argued, too-strong and too-short-term incentives can actually be counterproductive in boosting innovation.

These ideas may seem like simple ones. Sadly, though, misconceptions obscure many of these truths. The discussion of innovation is frequently confused, full of assertions that do not stand up to the light of day. These distorted ideas often lead to poor decisions by corporate managers, investors, and public leaders alike.

A Hybrid Model

The sweet spot, I argue, is a hybrid between the systems that produced Motorola and Skype: initiatives that combine the best features of the corporate research laboratory and the venture-backed start-up. In that way, the powerful motivations and focused goals associated with venture capital can be preserved, while the limitations that circumscribe the effectiveness of this intermediary can be overcome. The path that gets us to this hybrid is firmly laid out by economics.

But this hybrid is more easily conceptualized than implemented. Over the last decade, corporations have launched a variety of initiatives to foment "open innovation," or the development of ideas from outside the corporate walls. Examples have included research alliances with smaller firms, the proliferation of prize schemes for new discoveries, and efforts to encourage the development of software by opening up key programs owned by corporations to outside programmers. But many of these efforts have failed, in large part because of corporations' unwillingness to offer significant enough rewards or to relinquish control.

The opportunities and challenges at work here can be illustrated by looking at the experiences of corporate venturing programs—initiatives that have frequently involved corporate employees investing in start-up firms, whether teams within the firm (internal corporate venturing) or in stand-alone start-ups. In theory, the resources and long time perspective of the large firm should make these programs particularly effective. And, as we will see in chapter 6, the track record has been quite positive overall. But certainly many of these efforts have stumbled, with similarly recurring problems.

Revisiting Xerox PARC

The failures of the Xerox Palo Alto Research Center (PARC) are well known and well documented.[6] But fewer know the story of their aftermath. It's well worth exploring as an example of how corporate venturing can go both right and wrong.

During the 1960s and 1970s, Xerox experimented with computers and more generally sought to invent the electronic office of the future.

In particular, its Palo Alto Research Center was remarkably successful in developing ingenious products that would fundamentally alter the nature of computing. The Ethernet, the graphical user interface (the basis of Apple Computer's and Microsoft's Windows software), the "mouse," and the laser printer were all originally developed at PARC.

The culmination of much of PARC's innovation was its development of the Alto, a very early personal computer. The Alto's first prototype was completed in 1973, and later versions were placed in the White House, Congress, and various companies and universities. Nonetheless, the Alto project was terminated in 1980: it was a victim of differences between its East Coast operations and West Coast computer people and the exit of discouraged engineers, many of whom headed off to found their own companies in search of more recognition and financial rewards. By 1995, the value of the companies spun off from Xerox exceeded the market capitalization of the parent firm itself; by 2000, the spin-offs were worth ten times more than Xerox.

The establishment of Xerox Technology Ventures (XTV), the firm's corporate venturing initiative, was driven by two events in 1988. First, several senior Xerox managers were involved in negotiating and approving a spin-off from Xerox, ParcPlace, which sought to commercialize an object-oriented programming language developed at PARC in the 1970s. The negotiation of these agreements proved to be protracted and painful, highlighting the difficulty that the company faced in dealing with these contingencies. More importantly, a book appeared in the same year documenting Xerox's failure to develop the personal computer, *Fumbling the Future*.[7] Stung by the description in the book, Xerox chairman David Kearns established a task force, with the mandate of preventing the repetition of such a failure to capitalize on Xerox innovations.

Based on the task force's recommendation, Chairman Kearns decided to pursue a corporate venture capital program. He agreed to commit $30 million to invest in promising technologies developed at Xerox. As he commented at the time, "XTV is a hedge against repeating missteps of the past."[8] He briefly considered the possibility of asking an established venture capital firm to jointly run the program with Xerox, but decided that the involvement of another party would introduce a formality that might hurt the fledgling venture.

Modeling XTV after venture organizations had several dimensions. The most obvious was the structure of the organization. Although this was a corporate division, rather than an independent partnership like most venture organizations, the XTV partners crafted an agreement with Xerox that resembled typical agreements between limited and general partners in venture funds.

The spinout process was clearly defined in the agreement both to ensure that disputes did not arise later on and to minimize the disruption to the organization. The XTV officials insisted on a formal procedure to avoid the ambiguity that had plagued earlier corporate ventures. The agreement made clear that the XTV partners had the flexibility to respond rapidly to investment opportunities, as independent venture capitalists typically do. They essentially had full autonomy when it came to monitoring, exiting, or liquidating companies. The partners were allowed to spend up to $2 million at any one time without getting permission from the corporation. For larger expenditures, they were required to obtain permission from XTV's governing board, which consisted of Xerox's chief executive officer, chief financial officer, and chief patent counsel. Similar to independent venture organizations (but unlike many corporate programs), the program also had a clear goal: to maximize return on investment. The XTV partners felt that the ambiguous goals of many of the 1970s corporate venture programs had been instrumental in their downfall.

Not only was the level of compensation analogous to that of the 20 percent "carried interest" that independent venture capitalists received, and the degree of autonomy similar, but XTV operated under the same ten-year time frame employed in the typical partnership agreement. Under certain conditions, however, Xerox could dissolve the partnership after five years.

The analogy to independent venture organizations also extended to the companies in which XTV invested. These were structured as separate legal entities, with their own boards and officers. XTV sought to recruit employees from other start-ups who were familiar with managing new enterprises. The typical CEO was hired from the outside, on the grounds that entrepreneurial skills, particularly in financial management, were unlikely to be found within a major corporation. Over the objections of Xerox's lawyers, XTV insisted that the employees receive options to buy real shares in the venture-backed companies, in line with traditional Silicon Valley practices. The partners believed that this approach would have a much greater psychological impact, as well as a cleaner capital structure to attract follow-on financings by outside investors.

The independence of management also extended to technological decision making in these companies. The traditional Xerox product—for instance, a copier—was designed so that it could be operated and serviced in almost any country in the world. This entailed not only constraints on how the product was engineered, but also the preparation of copious documentation in many languages. These XTV ventures, however, could produce products for "leading edge" users, who emphasized technological performance over extensive documentation.

Between 1988 and 1996, the organization invested in more than a dozen companies. These covered a gamut of technologies, mostly involving electronic publishing, document processing, electronic imaging, workstation and computer peripherals, software, and office automation. These not only successfully commercialized technology that was lying fallow in the organization, but also generated attractive financial returns.

One successful example of XTV's ability to catalyze the commercialization of technological discoveries was Documentum, which marketed an object-oriented document-management system. Xerox had undertaken a large number of projects in this area for over a decade prior to Documentum's founding, but had not shipped a product. After deciding this was a promising area, XTV recruited Howard Shao and John Newton, both former engineering executives at Ingress Corporation (a relational database manufacturer), to lead the technical effort.

Shao spent the first six months assessing the state of Xerox's knowledge in this area—including reviewing the several three-hundred-plus-page business plans prepared for earlier proposed (but never shipped) products—and assessing the market. He soon realized that while Xerox understood the nature of the technical problems, it had not grasped how to design a technologically appropriate solution. In particular, the Xerox business plans had proposed building document management systems for mainframe computers, rather than for networked personal computers (which were rapidly replacing mainframes at many organizations). With the help of the XTV officials, Shao and Newton led an effort to rapidly convert Xerox's accumulated knowledge in this area into a marketable product. Xerox's depth of understanding—as well as XTV's aggressive funding of the firm during the Gulf War period, when the willingness of both independent venture capitalists and the public markets to fund new technology-based firms abruptly declined—gave Documentum an impressive lead over its rivals.

Documentum went public in February 1996 with a market capitalization of $351 million. XTV was able to exit a number of other companies successfully, whether through an initial public offering, a merger with an outside firm, or a repurchase by Xerox (at a price determined through arms-length bargaining). A conservative calculation indicates that the $30 million fund generated capital gains of at least $175 million for Xerox.[9]

The same assumptions suggest a net internal rate of return for Xerox (i.e., after fees and incentive compensation) of at least 56 percent, which compares favorably to independent venture capital funds begun in 1989, which had a mean net return of 14.9 percent.[10] These calculations of Xerox's IRR do not include any ancillary benefits generated by this program for the corporation. For instance, some observers argued that high expected value projects that might otherwise not have been funded through traditional channels due to their high risk were increasingly funded during this period, apparently out of the fear that they would otherwise be funded by XTV and prove successful.

Despite these attractive returns, Xerox decided to terminate XTV in 1996, well before the end of its originally intended ten-year life.[11] The organization was replaced with a new one, Xerox New Enterprises (XNE), which did not seek to relinquish control of firms or to involve outside venture investors. The XNE business model called for a much greater integration of the new units with traditional business units. The autonomy offered to the XNE managers and their compensation schemes were much closer to those in a traditional corporate division. As such, XNE represented a departure from several of the key elements that the XTV staff believed—and academic evidence suggests— are critical to success, such as their considerable degree of autonomy and high-powered incentives.

What explains the decision to terminate this successful program? One concern was that by rewarding only those working on marginal technologies, the program distracted researchers from their critical job of designing better copiers and printers. But the most frequently offered explanation emphasizes the shock that corporate leaders felt regarding the compensation promised to the venture team. (In the original agreement, Xerox had promised to pay 20 percent of the capital gains to the individuals running the program.) The top brass doubtless never anticipated the size of their eventual obligations, which appear to have totaled at least $44 million. Once they understood that midlevel functionaries in an obscure initiative were going to become the highest-paid executives in the corporation, their enthusiasm for the program disappeared overnight. (Never mind that the only reason they were to be paid so much was because they had created so much value for Xerox!)

Despite the short life of XTV—and many other corporate venturing programs—these types of hybrid initiatives are not impossible to pull off. When one looks at many of the costly failures involving innovation hybrids, one can argue that their failure was preordained from day one. The design of these efforts was so compromised that success was very unlikely. Conversely, many of the successful hybrids have been characterized by thoughtful design and implementation. This is not a realm where success is an elusive, almost random outcome; through careful thought and planning, the best of the corporate and venture innovation models can be achieved.

Roads Not Taken

The hybrid model I'm advocating, as difficult as it might be to achieve, stands in sharp contrast to how we usually think about innovation. A frequently invoked depiction of innovation is the "pipeline model," which suggests that innovations result inevitably from more spending on basic research. If we just spend more on basic research, the argument goes, applied research, development, and ultimately new products and services will follow automatically. Such beliefs underlie many "crash" initiatives to develop new products, whether in the private or the public sector. This is the classic model of corporate R&D.

Yet the actual track record of these "research push" initiatives has been miserable. Perhaps the most memorable of these failures was the U.S. Synthetic Fuels Corporation, a federal entity created under

Jimmy Carter's leadership with the mandate to spend the equivalent of a quarter trillion of today's dollars to stimulate advanced energy research. This effort was widely regarded as a complete failure, and was abandoned in 1986. Despite this dubious legacy, many of the Obama administration's proposals to address global warming through new technologies seem to subscribe to exactly this view of the innovation process: the efforts in the controversial 2009 U.S. stimulus bill to accelerate innovation relating to clean technologies such as solar and wind energy are a prime example.

An alternative view of where innovations come from can be termed the "great man" theory. Breakthroughs, it is claimed, are all about visionaries, who are periodically visited by flashes of genius. Such a view has a long pedigree—think of Archimedes running naked through the streets of Syracuse after a bathtub soak led him to solve the problem of determining gold's purity—and certainly has some truth. Many of the biographers of technology industry leaders such as Bill Gates, Steve Jobs, and Larry Ellison attribute their success to a superhuman ability to spot innovations.

But this view, too, is incomplete. Correctly predicting the future is just the beginning, not the end, of the innovation process. The annals of technology are rife with individuals and firms that had a clear vision of where the future was going, yet somehow failed to cash in on these insights.

A dramatic illustration of these propositions is a series of advertisements created by AT&T in 1993.[12] These ads offered a glimpse into the future, predicting such innovations as the ability to access books remotely, obtain automated directions while driving, take part in a video conference from a laptop computer, and pay tolls electronically. Each ad ended with the tagline "and the company that will bring it to you . . . AT&T." And while all of the technologies depicted in the advertisements did come true, the share of the profits from these products accruing to the AT&T Corporation—now a subsidiary of one of the regional phone operating companies—is infinitesimal.

Ascending the Heights

Getting innovation right is vitally important. The aftermath of the "Great Recession" has led to an enormous appetite for growth across the world. And the need for growth leads inexorably to a hunger for innovation. These words are being written in the summer of 2011, when the headlines in the United States have been dominated by a polarized discussion over the debt limit. This rancorous debate—while not the first and surely not the last such discussion—has highlighted the grim economic calculus that the United States and other advanced developed nations are confronting. A combination of huge unfunded liabilities and a lack of economic growth suggest a considerably less prosperous future unless growth can be ignited.

For instance, in the United States today the level of existing public expenditures and ongoing deficits arguably push the outer limits of sustainability. And the official balance sheets, if anything, understate the problem: unfunded liabilities, not reflected on the government's official balance sheet, for the two main health care entitlement programs alone, Medicare and Medicaid, are estimated to total $58 trillion dollars.[13]

Nor do these problems show any sign of easing any time soon: the latest Congressional Budget Office estimates show the U.S. government running a deficit until at least 2080.[14] Meanwhile, persistently high unemployment drains the limited resources for social services and reduces the pool of individuals paying taxes into the system. The situation in other advanced economies is arguably even worse: the United Kingdom, Japan, and many nations in continental Europe face a toxic mixture of low growth, huge debts, and substantial unemployment.

The easiest remedy for these problems, of course, is simply to cut costs—whether wasteful public programs or too-ambitious entitlements such as public employee pensions. But this strategy—however necessary in some cases—is likely to be far less efficacious than the average "Tea Party" enthusiast would care to believe. As the experiences of a number of nations have illustrated, too-severe cutbacks may have the effect of actually perpetuating the downturns that caused the financial crisis to begin with, depressing growth and sapping consumer confidence.

This dilemma is not confined to the advanced industrial nations. Numerous emerging economies have come under enormous stresses, facing social unrest and outright revolution. In many cases, the unevenness of economic progress has been a key driver of discontent—even if some have benefited from growth, it has not been as substantial or as widely distributed as needed.

And this brings us to growth. Economic growth leads to more working people and fewer unemployed, more tax revenues, and a great easing of the pressures that nations are under. And there are essentially two ways to get such growth, at least in advanced economies that cannot simply imitate breakthroughs that have already taken place elsewhere. One is to simply add more inputs: having workers, for instance, retire at a later age or run plants later into the evening. But this strategy is essentially a game of diminishing returns: nations can go only so far in pushing people to work longer and harder. (Unless, that is, one is in Greece, where trombone players and pastry chefs have traditionally been able to retire as young as fifty due to the hazardous nature of their work.) Nor, in a world of diminishing resources, is it clear that simply producing more is desirable.

The other alternative is much more appealing: to get more out of the existing inputs through a process of innovation. Such innovations can take many forms, from novel goods and services to new production processes to improved ways of organizing and managing. Since the pioneering work of Moses Abramowitz and Robert Solow in the 1950s, we have understood that technological change is critical to economic growth: innovation has not just made our lives more comfortable and longer than those of our great-grandparents, but has made us richer as well.[15] Innumerable studies have documented the strong connection between new discoveries and economic prosperity across nations and over time. This relationship is particularly strong in advanced nations—that is, countries that cannot rely on copying others or on a rapidly increasing population to spur growth.

We live in an age where technological innovation is more critical than ever for other reasons as well. Consider, for instance, the dismayingly long list of challenges that the world is likely to face over the next decades: environmental degradation, global warming, proliferating pandemics, terrorism, and the like. Our ability to respond to these challenges—many of which will require scientific and engineering breakthroughs—will be critical to our future and that of our children.

Innovation is complex and multifaceted, resisting easy characterizations. But an abundance of evidence suggests that greater attention to the ways in which organizations and incentives can shape the innovation process can produce significantly better results. That's what the hybrid model I'm presenting here aims to do. It isn't easy, as the example of XTV demonstrates, but it is something that, done right, will prove more effective than either of the two traditional ways that innovation has been pursued.

But to understand how to combine those two traditional models into a powerful system for consistently and efficiently producing new ideas, we have to first understand how each works as well as their virtues and drawbacks. Part 1 examines the "corporate model" in greater depth, starting with the fascinating story of where R&D came from in the first place.

Part One

The Traditional Model

2

Where R&D Came From

I N 1624, THE Latin edition of the British scientist, statesman, and philosopher Francis Bacon's *New Atlantis* first appeared.[1] In it, he describes a visit to Bensalem, an imaginary advanced civilization somewhere in the Pacific Ocean, and articulates the rationale for corporate R&D. One of the key institutions of Bensalem is Salomon's House, a research institution that may be government owned but is definitely independently run.

In Salomon's House, a wide variety of specialized actors all contribute to the R&D enterprise. These include those who review research published elsewhere (whom he named "depredators"), those who undertake new experiments ("miners"), teams that seek to translate the experimental results into practical applications ("dowry–men"), and those who plan and execute more illuminating follow-on experiments based on the initial efforts ("lamps" and "inoculators"). In this centuries-old work lies a key rationale for R&D laboratories today: the ability of multiple specialists to tackle complex problems, with a parent organization providing a coordinating role for all these activities.

Bacon's work also highlights that strong rewards motivate innovators. He wrote: "Upon every invention of value, we erect a statue to the
inventor and give him a liberal and honorable reward."[2]

Bacon didn't create any R&D facilities of his own. But his work articulated many of the powerful reasons for these facilities, arguments
that became particularly compelling in the waning days of the nineteenth century and early days of the twentieth. These years saw the
widespread adoption of the corporate R&D laboratory, establishing
the template that dominates innovative activities around the world to
this day.

We've certainly seen remarkable growth in R&D over the past six
decades—the amount of spending on R&D in the United States has
increased more than tenfold, even after adjusting for inflation—but
the lion's share of R&D spending took place, and continues to occur,
in industrial laboratories, not in start-ups or with entrepreneurs, as
many might guess. This was true in 1953, when industry performed
70 percent of the research, and in 2008, when corporations undertook
74 percent of the spending.[3] While there have been changes over this
period—for instance, academic research has sharply surpassed that
performed by the government—the basic pattern has been more of
stasis than change.

There have been sharper changes in who *pays for* this research. Between 1953 and 1979, the largest funder of U.S. industrial research was
the federal government. Starting in 1980, the federally funded share
fell sharply. This decline was driven by the increasing importance of
industries such as pharmaceuticals and information technology, and
the declining relative importance of defense and space, where research had been powered by federal dollars. Today, as was the case in
the 1930s, about two-thirds of all U.S. research is funded by industry.

Research today, as during the twentieth century, is still dominated by the very largest firms.[4] Although the percentages have fallen

somewhat over time, firms with over ten thousand employees still represent over half the research spending. Those with fewer than five hundred employees, a traditional definition of small business, account for less than a fifth of the total expenditures. These same patterns hold internationally.

The impact of a few key industries is also apparent if we look at the top-performing R&D companies in the world.[5] Life sciences companies represent four of the top five companies on the list (Roche, Novartis, Pfizer, and Merck), and eight of the top twenty-five. Information technology (including Microsoft, the only non–life sciences company in the top five), autos, and consumer electronics dominate the rest of the top twenty-five. Despite the attention paid by media and policy makers to the increasing amount of R&D being done in emerging markets, the dominant role of large firms in Europe, Japan, and the United States is apparent. The highest-ranked emerging market firms are the Brazilian corporation Vale (ranked number 65) and Petrochina (number 69).

These patterns also hold when we look at aggregate R&D spending globally. The United States alone accounts for 35 percent of world R&D spending, towering above the amount invested by such nations as China, Germany, and Japan. Elsewhere, as in the United States, the private sector drives R&D. For instance, in Japan, business enterprises perform 76 percent of the R&D; in China, 73 percent; and across the European Union, 61 percent. The only exceptions to the rule that the private sector constitutes the majority of private sector spending are small nations such as Holland, Greece, and Poland.[6]

This data suggests a key question: why has the corporate lab been so central for innovation? Part of the answer lies in the sheer difficulty of developing new technologies. Many of the most important technological discoveries do not simply involve one expert in a given area working alone. Rather, they entail combining insights from a variety of disciplines. The process of innovation frequently draws together individuals from a variety of perspectives, who must share information and combine ideas.

Having a diverse team at these laboratories might have a plethora of benefits after the discovery is made as well. A varied group within a laboratory is likely to be more effective in identifying new applications for recently developed technologies, or complementary ideas developed elsewhere to purchase or license. Great inventors, as economic historian Nathan Rosenberg points out, seem to be as inept as everyone else in anticipating the uses of their inventions. Among his illustrations is Thomas Watson's statement in 1947 about the pioneering computer that his firm has just developed, the IBM SSEC: because a single device could solve all the problems in the world needing scientific calculations, there was unlikely to be any demand for producing multiple units of the computer. Another example is Thomas Edison's conclusion that the primary application of the phonograph would be to record the deathbed wishes of wealthy gentlemen.[7]

If these kinds of discoveries really do need multiple experts, it makes sense to locate them in a single lab. Having them dispersed would raise too many complicating issues because trade in ideas among free agents is a challenging matter. Nobel Prize–winning economist Kenneth Arrow, in a 1962 essay, highlighted the essential problem: an idea about a new innovation, unless protected by a patent or other legal means, can be readily taken and used by competitors, thereby rendering it much less valuable.[8] As a result, an inventor seeking to sell a new idea faces a real dilemma. Unless he or she reveals key details about the invention, no one is likely to offer a substantial payment for the idea. But once the details of the breakthrough are revealed, the potential buyer has every incentive to express a lack of interest, and then exploit the idea illicitly. Although inventors attempt to limit these problems by requiring potential buyers to sign confidentiality agreements, these documents frequently prove ineffective, as the long and sad list of lawsuits between inventors and potential licensees illustrates. By bringing all the experts together under the umbrella of a single organization, the free flow of ideas can be greatly enhanced.

Another reason to bring the scientists together is the benefit of proximity in encouraging knowledge flows. Ever since the pioneering 1890 economics textbook of Alfred Marshall, there has been an appreciation of the benefit derived from locating innovators near each other.[9] In particular, he highlighted the importance of "knowledge spillovers": transfers of insights in a myriad of different ways between nearby researchers. This insight has been corroborated in many subsequent studies, which have highlighted the importance of inventors being near to each other. For instance, a detailed study of the laboratory floor plans and e-mail networks of a leading biotechnology company by Chris Liu, a strategy professor, led to the conclusion that being located near researchers in complementary but different fields on the same floor led to significantly more productive interactions.[10] By mixing and matching scientists and engineers from different fields, research labs should be able to maximize their productivity.

The Real Beginning

However prescient Bacon's early-seventeenth-century vision of the research laboratory, and however compelling the arguments, it took centuries for this institution to emerge. Inventors in the centuries after the publication of the *New Atlantis* continued to invent independently, seeking to commercialize their ideas through a combination of entrepreneurial ventures and licensing deals.

A typical path to the development and commercialization of innovation in that era was that of Elias Howe, the now largely forgotten (except by the Beatles, who dedicated the movie *Help!* to him) father of the sewing machine.[11] Born in rural Massachusetts in 1819, Howe had apprenticed to a textile mill at age sixteen. After the panic of 1837 cost him his job, he moved to Boston, but his frail health made it difficult to hold down a job. Lying in bed at home, he watched his wife sew to pay the family's bills. During this downtime, he conceptualized an idea for a sewing machine that would duplicate the action of the human hand and arm. In particular, his design combined three features that have characterized sewing machines ever since: a needle with the eye at the point, a shuttle operating beneath the cloth, and an automatic feed. He was awarded U.S. Patent 4750 in 1846 for this design.

The idea of a sewing machine was not new: people had been trying for at least half a century to create such a device. Earlier efforts had been star-crossed. Barthélemy Thimonnier never was able to find additional backers after his pioneering Parisian sewing machine factory was ransacked and burned in 1831 by the local tailors' guild, whose members feared that the new machines would render their skills obsolete.

Howe himself faced many obstacles during the commercialization process. After his first workshop burned down (no union members were to blame for that conflagration), he struggled to make a machine that cost less than $300, a huge sum at the time. Eventually, he moved to London to team up with a British garment manufacturer. But a falling-out left his family stranded overseas and essentially penniless. Upon returning to the United States in 1849, he launched a multiyear patent battle—financed by a mortgage on his father's farm—against Isaac Singer and Walter Hunt, who had introduced a highly successful machine that was ultimately found to have infringed Howe's patent.

His patent was upheld in 1854, and Singer was ordered to pay many thousands in back royalties. Soon after, the sewing machine manufacturers established a patent pool, and Howe negotiated a $5 royalty for each machine sold in America and $1 for every one sold abroad. This transaction made him a wealthy man, and allowed him to found Howe

Sewing Machine Company. But reflecting both his poor health and the arduous struggle of the previous decades, he died in 1867, at the age of forty-eight.

In light of Howe's story, the corporate lab certainly seems like a good alternative to the lone heroic inventor struggling through the wilderness alone. Three distinct factors can be seen behind the slow emergence of corporate labs, despite the compelling rationale for their existence. The first was the relative simplicity of technology prior to the late nineteenth century. Essentially, most cutting-edge technologies did not require great scientific expertise to develop, which meant that the need for the kind of specialization that Bacon discussed was modest. This is not to say that these early inventions were easy: few of us could design a successful steam engine. But much of the expertise for successful innovation in that earlier era appears to have been rooted in technical know-how, acquired largely through practice.

This simplicity of innovation in that era can be seen in the work of Naomi Lamoreaux and Ken Sokoloff, economic historians who have (along with Zorina Khan) examined nineteenth-century inventors, particularly those they term "great inventors," or individuals whose technological discoveries were notable enough to earn them inclusion in the *Dictionary of American Biography*. They showed that of all the patents awarded to great inventors born between 1820 and 1845, close to 80 percent went to people who had never gone to college. Those whom we might think of as the most natural inventors—educated types who had studied science, engineering, or medicine in college—accounted for only 10 percent of the inventions. It is not until the group born in 1845 and after—those doing the bulk of their inventing in the final decades of the nineteenth century—that technically trained inventors begin to dominate the ranks.[12] This technological simplicity also often went hand in hand with relatively modest costs, which made it easier for individuals to work alone.

Another way to illustrate the simplicity of technological innovation is to look at the first U.S. patent law. When this law was enacted in 1790, Congress saw fit to leave the review of patents to a commission of three men, most notably "first administrator" (and secretary of state) Thomas Jefferson. This group was responsible for determining whether the inventions were "sufficiently useful and important" without any staff of technology experts. It was felt that three well-trained generalists could adequately grapple with the entire range of technologies.[13]

A second factor precluding the formation of centralized research facilities was the nature of the American market. In the first half of the nineteenth century, the great distances and substantial cost of transport meant that it was far more efficient for most products to be made by separate firms for each regional market. It was not until the railroad boom after the Civil War that a national market emerged. Until this change, inventors often licensed their patents separately in each geographic region: Lamoreaux and Sokoloff found that among the patent licensing agreements filed with the U.S. Patent Office, the share representing distinct assignments to particular geographies fell from 23 percent in 1870–1871 to 5 percent in 1890–1891 to only 1 percent in 1910–1911.[14]

A final consideration was the unappealing nature of patents, at least in the United States and the United Kingdom. Between 1793 and 1836, the United States operated under a registration system, where any application meeting minimal technical requirements would get a patent. This led to a confused and disorganized morass of overlapping claims. Even in the years after this reform, the office was chronically understaffed and struggled to adequately review the applications coming in the door. In the United Kingdom, it was not until 1883 that an office resembling a modern one, with a formal process for ascertaining whether the invention was a reasonable one, was established.

Moreover, considerable ambiguity in U.S. patent law surrounded the question of whether an employer could take title to an employee's patent, reflecting the fact that the founders had focused on a vision of the inventor as a heroic individual genius.[15]

Over the second half of the nineteenth century, all three factors shifted: technology became more complex and costly to develop, transportation made markets more centralized, and the ambiguities around the value and assignment of patents were resolved. As a result, more and more inventors shifted from working for themselves to being part of a corporate research effort. This can be seen among the patents assigned to Lamoreaux and Sokoloff's great inventors. Among those born between 1820 and 1839, only 1.5 percent of their patents were assigned to a large industrial corporation. For those born in the next two decades, the fraction climbs to nearly 12 percent. Among those in the birth cohort between 1860 and 1885, fully 27 percent of their patents were assigned to large firms.[16] The age of the modern industrial research laboratory had been born.

The Modern Research Laboratory

Who gets the honor of being designated the first industrial research laboratory remains a matter of debate. Should it go to the Pennsylvania Railroad, who in 1875 hired a PhD chemist to address such questions as the proper viscosity of lubrication oil and the composition of steel for rails? Or to Thomas Edison and Alexander Graham Bell, who established "invention factories" in 1876 in Menlo Park, New Jersey, and Boston, respectively? Many historians of technology would point to the laboratories established by German pharmaceutical and dye manufacturers in the 1880s and 1890s as the first truly institutionalized research laboratories, and to General Electric's 1900 laboratory as the pioneer in America.[17]

Whoever is right, the key point is this: by the early days of the twentieth century, thoughtful industrial leaders recognized the need for the corporate research laboratory. The growth of these institutions was particularly rapid in the first half of the twentieth century.[18] The years between 1933 and 1946 saw even faster growth, particularly in sectors such as chemicals (including pharmaceuticals), electrical machinery, and transportation equipment.

The central corporate R&D laboratory was a dominant feature of the innovation landscape for most of the twentieth century, employing tens, and then hundreds, of thousands of researchers, many of whom were free to pursue fundamental science with little direct commercial applicability in these campus-like settings.

The most famous of these was undoubtedly Bell Labs, long considered one of the crown jewels of research worldwide. Funded for generations by AT&T's profits from its long-distance monopoly, its researchers made some of the most fundamental breakthroughs of the twentieth century, playing key roles in the invention of the transistor, laser, and communications satellite. Other breakthroughs—perhaps less immediately useful, but eventually far reaching—included radio astronomy, information theory, and the UNIX operating system. Its researchers were recognized with eleven Nobel Prizes, both for inventions like those delineated above and for more fundamental discoveries in quantum physics. IBM Central Research, with five Nobel laureates, was another leading exemplar.

Yet the process of creating these facilities, and defining their missions, was not painless. It included numerous false starts, as firms found that particular designs were not as conducive to promoting innovation as originally thought.

Consider the experience of DuPont, whose evolving research laboratories have been exhaustively documented by David Hounshell and John Smith.[19] In particular, after establishing its first laboratory in

1902, the firm struggled with three key questions: the management of research projects, especially the painful decision of when to kill a project; the retention of innovators in the research labs; and the appropriate organization of the research effort.

The first of these challenges centered around the decision of when to terminate projects. For instance, one high-profile early R&D project was to develop a continuous process for manufacturing black powder for explosives and weapons. Within a few years, however, the project was becoming more expensive than the executive committee had forecast and progress was slowing. Despite countless technical difficulties and an increasing risk of failure, the executive committee continued to appropriate hundreds of thousands of dollars to the project before it was ultimately abandoned in 1911.

Another example was the development of Stabillite. In 1904, DuPont's executive committee spent over $120,000 to license the right to this smokeless powder. By 1910, after appropriating nearly $300,000 for research, DuPont could not remedy its serious defects, including such issues as toxicity and a tendency to spontaneously combust, and the committee deemed Stabillite a commercial failure.

Some argued that such dry holes were a necessary by-product of the pursuit of visionary work; others argued that such blunders could be avoided with judicious management. For instance, A. J. Moxham, the R&D head at the time, noted: "The tendency of all experimenters is to continue to work on with an experiment just as long as any improvement seems to them possible. So tempting is this tendency that sooner than give up an experiment at a time when it has already reached the stage of profit they will retain control of it . . . because of the very love of the study they are making."[20]

Over time, the executive committee members realized that deciding when to abandon a project was a critical part of their role. To rely too extensively on signals from obsessed inventors would be a critical mistake. Rather, they themselves needed to more tightly monitor the evolution of projects, as well as encourage the development of a new generation of specialized managers of R&D projects.

A second challenge associated with the early years of the DuPont labs was employee turnover. From Pierre du Pont on down, the firm had emphasized the importance of continuity among the scientists and engineers at the research facilities. But the practical realities of managing innovative technologies in a complex organization cut against the achievement of this target. In particular, an enduring challenge in the divisional laboratories was that promising technologists would be swept along with their technologies when it came time to move from the bench to the factory floor. While such transfers were completely understandable—after all, who had a better understanding of nascent technologies than the individuals who developed them?—in the longer term, this practice served to deprive the labs of some of their most talented members.

But surely the most vexing issue DuPont faced was the extent to which research should be closely linked to the various operating divisions, or instead, centralized. Over the first decades of the twentieth century, DuPont pursued industrial research through two separate in-house laboratories. The Eastern Laboratory pursued short-term projects, which could rapidly be put into practice, while the Experimental Station pursued longer-term riskier projects. The decision to take these two disparate routes was based on a sense that each strategy would have unique strengths: the former laboratory would respond to the divisions' needs in a timely manner and have a well-defined mission, while the central R&D lab would limit duplication and pursue projects that could eventually benefit the entire company.

The inability to quantify the benefits from the central research effort proved to be a major challenge. The Eastern Laboratory could point to tangible indications of its success: the actual dollar savings garnered from its research into topics such as nitroglycerin manufacture and low-freezing Lydol dynamite. Defenders of the Experimental Station were hard pressed to come up with similar examples: Charles Lee Reese, for instance, asserted that the central lab "should not be held accountable for poor return-on-investment showings for such research work" since it brought "enormous, though intangible, rewards" to DuPont.[21] Adding tension to these dynamics was the desire of many of the founders of the central laboratory—who often were recruited from universities—to retain their status within the academic community by recruiting the best and brightest minds, however removed from practice.[22]

In the aftermath of the recession after World War I, DuPont shifted to a decentralized R&D strategy as part of a broader corporate restructuring. This reflected a sense that the central R&D effort had been insufficiently responsive to the day-to-day requirements, and instead focused on too-academic work removed from the firm's pressing needs. Instead, the firm shifted to an approach where each division would have dedicated staff groups, each meeting its product family's needs, including R&D. While these divisional laboratories initially focused on incremental research, over time, their role expanded to encompass the development of entirely new products.

By the late 1920s, the pendulum swung again, and DuPont shifted to a model with a central research organization that coordinated the various operating divisions' R&D units. Motivating this change was a sense that without a central R&D unit, long-term projects were either not pursued or undertaken ineffectively. Charles Stine, who led the effort to revive the central research function, argued that there were four critical benefits that could accrue from a central research effort:

1. Gain goodwill from consistent publishing of academic papers

2. Attract more PhD chemists to the company

3. Allow for information bartering from other research laboratories

4. Possibly give rise to commercial innovations, particularly of a more radical nature[23]

These arguments carried the day, and as part of his 1927 budget, Stine was allocated $115,000 to build a new laboratory for basic research, which the researchers soon dubbed "Purity Hall."

But the dramatic speed and manner with which this shift to basic research bore fruit must have surprised even Stine. In 1928, DuPont hired a brilliant but mercurial chemist from Harvard, Wallace Carothers. Even as a graduate student, Carothers had tired of the grind of teaching and other routine academic tasks, complaining, "It contains all the elements of adventure and enterprise which a nut screwer in a Ford factory must feel on setting out for work."[24] DuPont persuaded him that at the corporate laboratory he could focus on what he loved—research—and put all the odious other activities behind him. Within two years, Carothers and his colleagues had produced a series of pioneering articles on polymers, documenting the basic ways in which these macromolecules could be created and manipulated. Within a few more years, his laboratory had produced the predecessors of neoprene (artificial rubber) and nylon. Before its division was sold to Koch Industries in 2003, nylon was estimated to have accounted for as much as $25 billion in revenues for DuPont: a pretty good return for an estimated total R&D investment of $4 million.

The decades leading up to World War II saw an extraordinary surge in the adoption of corporate laboratories. While in some cases, the decision to set up such a laboratory may have been driven by a desire to get on board with a management "fad," for many firms, these facilities proved to be attractive investments. Nonetheless, the effective design and management of these facilities proved to be an ongoing challenge, which would only intensify after World War II.

The Postwar Era: Growth and Reconsideration

The victory in World War II ushered in a period of unprecedented interest in basic science.[25] In the aftermath of the Manhattan Project to construct the atom bomb, government funding of science—and particularly basic science—in the United States increased dramatically. Much of the growth was in the defense arena, but the new National Science Foundation also played a key role.

This shift was mirrored in the corporate sector. Company after company established laboratories devoted to basic science, from IBM's Watson Scientific Computing Laboratory in 1945 to the Scientific Laboratory at Ford's Dearborn Engineering Laboratory in 1951. In other cases, such as DuPont, existing basic research efforts were dramatically expanded.

These initiatives seem grounded in the "pipeline model": the supposition that spending on basic research would automatically translate into valuable commercial insights. Not surprisingly, in many cases this assumption proved to be optimistic. For instance, at DuPont, the push for more basic research resulted in having each of the eleven operating divisions undertake fundamental R&D initiatives of its own. Not only were these divisional labs typically substantially less successful than the central lab in undertaking basic research, but the effect on the central laboratory was also deleterious. To minimize duplication, the central unit moved to work that was very distant in its applicability and whose relevance to the firm's technical needs or skills was quite minimal. The unsurprising consequence was that in the decades that followed, an increasing number of questions were raised, both within and outside the firm, as to whether DuPont's central research effort was still a valuable asset for the corporation.

By the late 1960s, and accelerating in the ensuing decades, companies began rethinking their commitments to large R&D expenditures, as they sought to take stock of their central lab's achievements in the postwar era. As Hounshell put it, "DuPont had no new nylons. Kodak had no radically new system of photography."[26] (As we know with the benefit of hindsight, Kodak's researchers understood the promise of digital photography early on. But the firm went about commercializing its pioneering digital camera in a highly problematic manner; for instance, marketing "film-based digital imaging" rather than purely digital cameras.[27]) Even at the laboratories that were most successful from a scientific viewpoint, such as at IBM, questions regarding the returns from these substantial investments were increasingly being raised: these facilities were often seen as producing top-class science, but having limited relevance to the firms' commercial needs. And more specifically, critics wondered about the effectiveness with which research laboratories were organized and managed. In a prescient 1964 essay, Joseph Bailey wrote, "We are largely fumbling in the dark as we try to insure [researchers'] motivation, or to 'manage' the creative talents we have hired them to contribute."[28]

Taking Stock

The late 1980s and 1990s saw passionate discussions around the need for and performance of corporate R&D. Not surprisingly, on the part of the scientific and engineering communities, there was substantial handwringing and concern. The National Science Board wrote in a high-profile report: "The United States faces an emerging risk of losing its traditional strength in pioneering discoveries and inventions. Most pioneering inventions of the past that have created the basis for new industries have originated in either corporate laboratories of private firms or in research universities. Both these institutions are under severe stress today."[29] Contemporaneous quotations from dozens more essays published in the United States and Europe bemoaned these cutbacks. The diagnoses offered ranged from excessively short-term managers to the deleterious effects of pressures from hedge and mutual fund investors.

But a brutal fact remains: most of these firms were simply not very good at doing research during this period, or at least in translating that research into profits. Two influential studies published around this time made the point compellingly.

The simpler approach was by financial economist Mike Jensen, who simply compared the stock of firms' R&D and new capital expenditures to their market value.[30] Essentially, this analysis is asking how much of the market value of firms can be accounted for by these two critical investments. We would expect in most cases, this would be a relatively minor fraction: after all, these firms had many assets other than their knowledge, such as strong brands built up via advertising, well-developed distribution networks, and dedicated and specialized sales forces. None of these other sources of value are captured by the stock of R&D and net capital expenditures.

When Jensen computed these values at the end of 1990, the answer was very different for some firms. For General Motors, R&D and net capital expenditures between 1980 and 1990 totaled $62.8 billion, yet the market value of its equity was only $20.8 billion. For Xerox, the comparable numbers were $8.6 and $3.2 billion. And so on, for a rogue's gallery of American firms. Even if a firm's other assets were useless, and the spending on research and new equipment prior to 1980 had no remaining value, these investments over the past decade had clearly destroyed value. In these cases, the firm's valuation did not reflect even these expenditures, much less the many other assets that the corporation controlled.

A more sophisticated approach to this question was taken in a high-profile 1993 analysis by economist Bronwyn Hall.[31] In this analysis, she sought to compute the return for firms from their investments in R&D and compare it to the return from more traditional capital expenditures (e.g., in new machine tools or factories). She found that during the 1970s, the ratio of returns from investments in R&D and traditional investments was about one. This state of affairs is what economists would anticipate: if one form of investment was dramatically better, the firms would switch to spending funds on the other category.

But in the 1980s and early 1990s, the returns to R&D relative to other investments fell sharply. A dollar of spending on R&D was enhancing market value only a quarter as much as a dollar's investment in traditional assets. These findings suggested that the market had become deeply disillusioned with corporate R&D as a route to creating value.

Final Thoughts

The story of the rise and fall of corporate R&D underscores three key themes.

1. There was a compelling rationale for establishing centralized research facilities, which appears to have been as true at the end of the twentieth century as at its beginning.

2. Many companies struggled to define the appropriate scale for their organizations, with numerous reversals of course and false starts.

3. The closing years of the decade saw a pulling back from many of the most ambitious visions as to what these facilities could accomplish in the face of repeated disappointments.

The next chapter examines the developments of the past two decades, especially the organization of these facilities and the nature of the rewards for researchers.

3

The Changing Face of Corporate R&D

I N 2006, NETFLIX, the video rental and streaming service provider, offered a prize of $1 million to the first team that could improve upon its movie recommendation algorithm by a set amount.[1] Up to that point, Netflix had relied on its Cinematch recommendation system to suggest movies. This system analyzed the accumulated ratings (on a five-star scale) of users to make personalized suggestions based on their tastes. Cinematch's performance was rated by comparing the quality of the model's predictions of a subscriber's taste and the actual rating made by the viewer after she saw a movie (more technically, by computing the root mean squared error, or RMSE, of the predicted ratings). Over time, internal enhancements of the algorithm and system performance had resulted in a 10 percent improvement in the RMSE, but Netflix, confident that further progress could be made, sought to use the Netflix Prize to encourage external research.

To assist participants, Netflix released a scrubbed, anonymized dataset of over 100 million of its movie ratings. It withheld over

3 million of the most recent ratings from those subscribers and challenged contestants to make predictions for those 3 million. Submissions were evaluated by computing the RMSE for a submitted set, and the results were posted to the leaderboard so that other contestants could see their standings at all times. Also, Netflix required the competition leaders to publish their techniques publicly so that everyone could benefit from the results. In addition to the $1 million prize to the winners, it gave away various "progress prizes" of $50,000 each. These progress prizes were awarded every year for the top result to date, as long as the leader's algorithm improved upon the RMSE of the previous prize winner (or of Cinematch in the first year) by at least 1 percent. The annual reward inspired continual efforts to surpass other contestants, as did the real-time leaderboard.

One of the intriguing aspects of the Netflix Prize competition was the firm's encouragement of collaboration between teams, mimicking the collaborations within an industrial research lab. After independent teams saw their progress slow during the second year, they began to combine their approaches. Teams seemed to derive the greatest benefit from far-out ideas—ideas that would fail to predict movie preferences if used alone—especially toward the end of the competition.

In fact, the seven members of team "BellKor's Pragmatic Chaos," which won the prize, came from three merged teams. The collaborators were successful, despite the fact that they never met each other face-to-face before receiving their award in September 2009. The group prevailed after a three-year competition, one that included submissions from more than forty thousand teams from 186 countries.

The Netflix Prize represents two of the major trends in corporate R&D over the past couple of decades: the decisive shift away from centralization, and the move toward more focused incentives to more closely link rewards to performance.

Away from the Center

The first of these shifts—the movement away from central research facilities—stems from the challenge of managing complex portfolios of research projects. Making judgments about when to continue and when to cut off research endeavors can be daunting in the best of circumstances. And when one has a very large number of projects in laboratories around the world, as characterizes many modern research-intensive large corporations, the challenge can be overwhelming. The odds are that numerous promising technologies will languish undeveloped and unappreciated.

The danger that potentially major technologies might sit underutilized in a central laboratory can be illustrated by considering the origins of a firm that today goes by the name of Zumobi.[2] In 1999, John SanGiovanni joined Microsoft's Redmond central research laboratory with a mandate to improve user interfaces for smartphone devices. John's background had not been that of traditional technologist: his prior positions had been in the entertainment division at Walt Disney Studios and as part of the Advanced Technology Learning Solutions team at Michael Milken's for-profit learning company, Knowledge Universe. During these experiences, he had become fascinated with the ways in which the then-nascent smartphone technology could transform the ways that individuals access and use information.

In 2003, SanGiovanni began working at Microsoft on smartphones as a consumer experience. In conjunction with outside academics, notably Dr. Ben Bederson of the University of Maryland, he developed a mobile software platform, originally called LaunchTile. The program enabled individual users to access, download, and deploy various applications (which they called "tiles") on a mobile phone. The technology featured a zooming user interface, through which users could readily navigate across the tiles by pressing a touchscreen. As a trade magazine described it, "The tiles themselves—which can be anything from news, to traffic, to mini-games, to whatever—are developed with an open API [application programming interface], encouraging developers large and small to get involved . . . The concept sorta has to be seen to be fully appreciated."[3] This technology led to a patent filing in February 2005, almost two and a half years before Apple's release of the iPhone.

Given this early head start, why is it that Apple and not Microsoft went on to garner the lion's share of the profits from the smartphone market? As SanGiovanni socialized the idea in the firm, the leadership of the mobile product group at Microsoft saw the promise of the new technology. But at the time, the firm was committed to the Windows Mobile platform for cellular devices. As SanGiovanni related, "It was hard to get them to adopt it. They had their own platform and a lot of constituencies to satisfy. Integrating the Zumobi platform with [Windows] Mobile would require at least thirty people and $10 million—and that wasn't going to happen."[4] Ultimately, Microsoft spun Zumobi out of its research laboratory as a freestanding firm, which was funded by the venture capitalists Oak Investment Partners and Hunt Ventures. By the time Zumobi was established as an independent entity, however, the iPhone had begun acquiring momentum, and the window for a new platform appeared to be shutting. Zumobi has subsequently transformed itself into one of the leading premium application developers for all mobile platforms, including Apple's App Store.

The fate of the Zumobi platform is particularly poignant given the subsequent evolution of the Windows Mobile software it unsuccessfully tried to displace. Microsoft largely abandoned the software platform in 2010 in favor of the incompatible Windows Phone operating system, as a response to huge losses of market share to the Android and iPhone systems.

In reaction to stories such as this one, firms have responded by streamlining and decentralizing their research. Many corporations have undertaken a fundamental reexamination of the mixture of activities between their divisional and centralized research facilities. Reflecting both a perception of disappointing commercial returns and intensified competitive pressures, firms have deemphasized central research facilities in favor of divisional laboratories.

The share of internally funded research supported by corporate headquarters (which typically went to fund central labs) fell steadily during the 1990s, while the corresponding funding by divisions for their own laboratories rose sharply.[5] Corporate spending on basic research, an activity typically associated with central R&D facilities, fell steadily during these years.

These statistics, however, do not really convey the magnitude of the transitions at many firms. In some cases, central R&D was simply cut. When Kodak replaced Kay Whitmore as CEO in the early 1990s, the board commented that a key failing was "spend[ing] too much on R&D without getting results." In the years afterwards, his successor, George Fisher, cut tangential projects in pharmaceuticals, household products, and medical tests, focusing the efforts on the firm's core film markets. Kodak's R&D spending fell from $1.6 billion in 1992 to $859 million in 1994.[6] By 2010, a much-reduced Kodak spent only $321 million on R&D, and in 2012, the firm filed for bankruptcy.

The most dramatic example of such cuts was doubtless the fate of Bell Labs.[7] This august institution is today a shadow of its old self. In 1996, AT&T—facing intense pressure from competitors and investors after telephone deregulation—made the fateful decision to spin out Bell Labs as part of a new entity called Lucent Technologies. The new entity almost immediately undertook a heady series of acquisitions, culminating with the 1999 purchase of Ascend Communications for $24 billion. These steps were soon revealed to be foolish, when the demand for telecommunications equipment fell sharply after the NASDAQ crash of 2000. Lucent's stock price rapidly fell from a high of $84 per share to less than $1.

Under intense pressure from the financial markets, Lucent cut the staff of Bell Laboratories from thirty thousand employees to fewer than ten thousand. The lab was further tarnished in 2002 when one of its key scientists, Jan Hendrik Schön, was found to have falsified data in more than a dozen papers. After a merger with the French giant Alcatel—which failed to stop the decline in the underlying business—the firm's research operations were further trimmed. By August 2008, only four scientists were still working in Bell's fundamental physics department, with a total research staff of fewer than a thousand employees.[8] A recent example of dramatic cutbacks was at Pfizer, the world's second-largest R&D-performing firm as late as 2010.[9]

But the changes at many other firms have been not so much a matter of making cuts but rather of fundamentally rearranging the way research laboratories work. One dramatic example was the changes at London-based GlaxoSmithKline (GSK) in 2008.[10] In this year, the company—one of the world's leading pharmaceutical firms—dramatically restructured its traditional slow-paced and bureaucratic system of R&D to emulate the relatively fast pace and entrepreneurial system of biotech companies. Discovery Performance Units (DPUs)—compact, autonomous, highly accountable, and extremely specialized teams consisting of seven to seventy-eight researchers—were established to manage the innovation process.

These changes were a reaction to a centralization that had characterized pharmaceutical R&D in recent decades. Moncrief Slaoui, GSK's chairman of research and development, said:

> In the 1980s, pharmaceutical companies saw how profitable new drugs could be. To increase profits, they tried to scale up R&D by industrializing it. We created huge departments where we gave scientists very small tasks that they could perfect doing and then pass the results on to the next scientists to do the next task because that is how we know to scale up industrial processes . . . I think this was a big mistake.[11]

By breaking the research efforts into smaller teams, GSK's management sought to encourage more collaborative thinking and interactions than the traditional centralized research facility would allow.

In setting up this program, the firm also instituted a much tougher review process for deciding which project to fund. In the new model, each small lab-scale DPU would "pitch" its program to a Discovery Investment Board, composed of both internal senior R&D executives and outsiders, including a banker, a leading scientist, a venture capitalist, and a biotech CEO. At the end of a three-month review process, successful DPUs would be awarded three years' investment money, a sum that often totaled hundreds of millions of dollars. In the ensuing years, the review board would track each unit's progress against various performance and value-creation metrics and could terminate funds if progress was lagging. As Slaoui noted: "In the past, your budget was typically last year's budget plus a little more. You got your money and did whatever you did. The new system will eliminate any notion that you are entitled to your money."[12] A related change was to push the DPUs to rely heavily on outside resources, essentially creating competition between GSK's internal units specializing in formulation of new drugs, animal studies, and the like, and external service providers. While the long lead times associated with drug development mean that a definitive assessment of these changes can only be made in the future, preliminary indications are promising.

The Rise of Open Source

An even more extreme form of decentralization has been the decision of firms to "open up" their technology for the world to work on. In the past decade, this trend has been seen most dramatically in the software industry, where companies have frequently released proprietary code using an open source structure to guide its further development.[13] In many cases, not only does the firm promise to keep the software available for all users, but anyone who makes use of the material must agree to make all enhancements to the original code available under these same conditions. For example, IBM released half a million lines of its Cloudscape program, a simple database that resides inside a software application instead of as a full-fledged database program, as well as its Eclipse software development environment. Another illustration of this trend is Hewlett-Packard, which released webOS in December 2011 as an open source project: earlier in the year, the firm had been intending to use the program as its next-generation operating system for all its computers and devices.[14]

Of course, these strategies are not motivated solely by the desire to benefit mankind: such actions are comparable to giving away the razor (the code) to sell more razor blades (the related consulting services that IBM and HP hope to provide). Not surprisingly, firms have generally sought to release proprietary code under an open source license when they anticipate that the increase in profits from the related segments will offset any profits lost by converting the software to open source. The temptation to go open source is particularly strong when a product is lagging behind the market leader, but the firm sees a possibility that widespread use and further development will increase the profitability of the other products or services.

Numerous challenges appear when a for-profit firm seeks to undertake such an extreme form of decentralization. Potential users are likely to fear that the commercial entity may not be serious when promising to share the benefits from the project, and consequently may not be willing to make contributions. In particular, a corporation may not be able to credibly promise to keep all source code in the public domain forever, especially if the user contributions take the project in a direction that is counter to the specific technological approach being promoted by the firm.

These difficulties help explain why Hewlett-Packard released much code through Collab.Net, a venture by leading open source programmers, which organizes open source projects for corporations that wish to open up part of their software. In effect, Collab.Net offers a kind of stamp of approval, or certification, that the firm is committed to the open source project: the ongoing involvement of open source developers was assured by Hewlett-Packard through the employment of visible open source developers and the involvement of O'Reilly, a technical book publisher with strong ties to the open source community. The Apache Software Foundation, the well-established open source project that IBM asked to manage its Cloudscape program, played a similar role in that case. While certainly these strategies have attracted a lot of attention in executive suites and press rooms, it is less certain that they have generated extensive profits for firms to date (though detailed information on profitability is very hard to come by!).

These are just a few of the approaches seen in recent years. The increased reliance on strategic alliances, competitions, and divisional laboratories are other examples. But whatever the variations pursued, the overall trend is clear: the era of the primacy of the central research lab appears to be over.

Taking Stock of Decentralization

Economists have offered several theories for how the broad scope of the central research laboratory might affect performance.

The work of Jeremy Stein, for example, suggests that the broad range of technologies considered in a diversified research lab can also contribute importantly to efficient decision making.[15] When investment opportunities are poor in one technological area, he argues, managers can maintain their overall capital budget (which they value in and of itself) while still making good investments in their other sectors. By contrast, managers of narrowly focused divisional labs with poor investment opportunities have no place else to invest and, in an effort to maintain their capital budget, may end up investing in unprofitable projects. Stein's model suggests that when investment opportunities are poor in one sector (say, life sciences), diversified central labs will be more prone than specialist firms to scale back investments in that sector and scale up their investments in more promising sectors (say, communications).

One of the critical elements of the Stein model is that the CEO—who has no vested interest in making investments in any particular sector— gets to decide where capital is allocated. But these decision rights are not as clear cut in large corporations. It may simply be too hard to change direction: chief technology executives sometimes refer to their task as steering an ocean liner. Another problem relates to the information that the senior managers receive. The executive suite is likely to depend critically on technologists to tell them where the greatest opportunities are. And these accounts are likely to be biased, especially given the passion and egotism that often characterize so many innovators. For instance, the "not invented here" syndrome may lead researchers to downplay competing technologies developed by other firms: whether motivated by overconfidence or self-interest (they want their own pet project to be funded), scientists and engineers are often dismissive of ideas developed elsewhere.[16] The advantage of a generalist form of organization may not be as great as suggested above. This prediction is in line with the view expressed in much of the literature about the failings of "internal capital markets," the authors of which (including, ironically, Jeremy Stein himself) argue that diversified firms have a difficult time redeploying capital into sectors with better investment opportunities.[17]

Economic theorists—perhaps unsurprisingly—can make a compelling case either way. What does the experience in the real world show? Probably the most dramatic way to study the importance of focus on the performance of research is to look at cases where firms change their structures.

Amit Seru examines cases where firms take over other concerns in an unrelated industry.[18] He looks at the research productivity of these firms before and after the deal, comparing them to a control group. The controls consist of firms that tried to undertake mergers, but whose deals fell through for some reason unrelated to the firms' R&D policies (for instance, where the deal was scratched because of a sharp drop in the overall stock market). He finds that the quality of the patents after the completed mergers falls sharply, measured through patent citations.[19] These results suggest that the internal workings of a conglomerate bring about a reduction in the quality of research conducted there. Even the same researchers who worked at the company before the acquisition become less productive after the transaction. Finding synergies across various innovative activities within a firm seems very hard.

Morten Sorensen, Per Stromberg, and I take the opposite approach, examining the changing innovations of firms after a private equity transaction, which typically separates one division of the firm from the rest of the operation.[20] Before the transaction, there are few differences between these companies and similar firms. But after the transaction, the bought-out firms seem to become considerably more innovative: although the number of patent filings does not change rapidly, the patents filed are more important (more frequently cited). The firms drop innovative projects in more marginal areas, and the quality of work in their core area increases.

Taken together, these results suggest the power of focus: firms undertaking too many projects in too many different areas may have a hard time discerning which areas are really relevant, and have a less successful research portfolio as a result. Meanwhile, knowledge spillovers, which the reader will recall was one of the important arguments for central research labs we sketched out in the previous chapter, seem less important than anticipated. These studies cannot tell us definitively whether firms would be more efficient with central-ized or divisional labs: to really answer this question, we would prob-ably need to run an experiment where we instruct random firms to switch their R&D lab structure, which would be a little hard to pull off! But the results suggest that there are reasons to be skeptical about the centralization of diverse research activities.

The Appearance of Incentives

Besides decentralization, the past twenty years have seen focused attention on the provision of incentives in corporate labs. Corpora-tions have traditionally downplayed the importance of monetary incentives for researchers. Typically, as part of the employment agree-ment signed on the first day of work, researchers sign over all rights to their ideas.[21]

Generally, these agreements are sweeping indeed, as David Barstow discovered.[22] From 1980 to 1994, Barstow, a computer scientist, worked for the oil services giant Schlumberger. At the start of his employment, he signed a standard invention assignment agreement, which included these key clauses:

> 3. Employee shall promptly furnish to Company a complete record of any and all technological ideas, inventions and improvements, whether patentable or not, which he, solely or jointly, may conceive, make or first disclose during the period of his employment with [Schlumberger].
>
> 4. Employee agrees to and does hereby grant and assign to Company or its nominee his entire right, title and interest in and to ideas, inventions and improvements coming within the scope of Paragraph 3:
>
> a) which relate in any way to the business or activities of [Schlumberger], or

b) which are suggested by or result from any task or work of Employee for [Schlumberger], or

c) which relate in any way to the business or activities of Affiliates of [Schlumberger],

together with any and all domestic and foreign patent rights in such ideas, inventions and improvements.

During this time, Barstow and his brother had a side project: a computer-based simulation of a baseball game. This effort led to four patents, two of which were filed and one of which was issued while Barstow worked for Schlumberger. Barstow had described the project to both his boss and the firm's general counsel for software, but neither saw a problem: after all, it was a long way from analyzing oil drilling data to fantasy baseball.

Fast-forward to 2004, a decade after Barstow had left the firm. The brothers' holding company sued Major League Baseball, arguing that some of the products offered on its website infringed their patents. The baseball solons turned around and went to Schlumberger and worked out a retroactive agreement to license all of the oil firm's intellectual property related to baseball simulations. Major League Baseball then moved for dismissal of the brothers' case, on the grounds that the sports league had a valid right to use all the patents, including the ones that had been filed after Barstow left Schlumberger. The brothers, on the other hand, argued they had clear ownership to the patents. On the basis of a very short hearing, the district court ruled that it was clear that both discoveries were "suggested by" and "related to" Barstow's work for Schlumberger. The fact that the work and the simulation both involved computer analyses of some type was sufficient for the judge. Although on appeal, the judge was ordered to undertake a more detailed investigation, whose findings favored the brothers, the case made clear the extent to which the assignment agreement that new employees sign every day limits their ownership of their ideas.

In almost all cases, the employers own new inventions. But what happens then? Typically, these ensuing dynamics are one-sided. Examples are rife of scientists and engineers who made major discoveries for their employers, yet got only token rewards. A canonical example is the $2 payment that Raytheon made to Percy Spencer as his sole compensation for discovering the microwave oven in 1945, a product he persisted in working on despite skepticism from his superiors.[23] Raytheon remained a major player in the production and operation of these ovens, which were introduced first in nuclear submarines, then more generally through the 1970s.

At least Spencer eventually became a senior vice president and had a Raytheon laboratory named for him in 1958. For a more depressing illustration of the lack of rewards for corporate researchers, consider the story of Shuji Nakamura.[24] As a lead research scientist at the midsized Japanese firm Nichia, he came up with a critical discovery in 1993: a blue light-emitting diode. These tiny, energy-efficient lights could be used in a wide variety of applications, from data storage to traffic lights. Combined with other primary colors, these diodes can produce white light, which may one day replace the traditional lightbulb. This insight allowed Nichia to expand its employment sixfold, and to earn profits estimated by the Tokyo District Court at $1.1 billion by the time the original patent governing the invention expired.

Nakamura's reward for this discovery was a bonus of $180. Despite the seeming inequity in this compensation, this would likely have been the end of the story had not the firm pushed its luck. When Nakamura was offered a post at the University of California at Santa Barbara in 1999, Nichia demanded that he sign an agreement not to do any research into light-emitting diodes for the next five years. When he refused, Nichia turned around and sued him. A year later, Nakamura himself struck back, alleging that his former employer had not given him a fair share of the profits from the blue LEDs. A key claim was that Nakamura had developed the blue LED on his own, after the company ordered him to suspend work on the project. Ultimately, the Japanese courts ordered Nichia to pay out $8 million to the inventor, in a ruling that was seen as precedent-shattering.

These anecdotes are backed up by large-sample studies as well. By and large, researchers have received modest pay for performance. For instance, Frederik Neumeyer presents numerous case studies of firms that either had abandoned their efforts to link substantial compensation to research discoveries or never had such schemes.[25] The view of Lawrence Hafstad, the vice president in charge of research at General Motors, was representative of the corporate attitudes at the profiled firms:

> I have been signing papers giving patent rights to some employe[r] or another since 1940 when I left work in pure science. I have never felt "forced" to sign these papers, as current liberal writings never miss an opportunity to imply . . . Then why did I sign? I believe the answer is that for me as for most engineers, I consider myself to be a professional problem solver. The problems may or may not involve the need or invention, but if they should, I feel as competent to invent as the next man . . . I personally have preferred to work as a problem solver on a salary, with invention being incidental, rather than as a free-lancer seeking riches from a single invention.

Similarly, the standard textbooks on the field inveighed against linking compensation to performance for inventors: for instance, a leading work published in 1990 stated: "In a research and development organization, indeed in most complex professional organizations, a number of reasons make tying pay inexorably to performance appraisal an imprudent approach."[26]

Larger firms, in particular, seemed to be considerably more reluctant to employ contingent compensation. For instance, Todd Zenger and Sergio Lazzarini surveyed 245 electrical engineers engaged in design and development in Boston and Silicon Valley in the early 1980s.[27] They found that those working at firms with fewer than three hundred employees were much more likely to have a higher ratio of bonus and awards to compensation from salary: an estimated 15 percent versus 6 percent for their larger counterparts. They were also considerably more likely to hold equity in their firm.

The skepticism about incentive compensation for researchers is understandable. The economists term the essential problem that makes designing incentive schemes in this setting difficult "multitasking."[28] An individual in a research laboratory is often asked to pursue multiple activities, some of which can be precisely measured while others cannot: his progress in generating patent filings may be easy to assess, but time spent helping others with their initiatives is impossible to monitor. If he is to be rewarded under a pay scheme with high-powered incentives, the researcher will face a temptation to focus on what can be measured (and rewarded). If, for instance, the compensation scheme leads the researcher to neglect collaborations with colleagues, the purpose of investing in the research laboratory is undermined.

The detrimental effects of such "multitasking" problems are captured in Leonard Sayles's account of a major industrial steel company's research laboratory in the early 1960s, which sought to link compensation to results.[29] To implement this idea, management required that to be eligible for bonuses, each group had to produce three technical reports per researcher employed. In short order, researchers began hoarding their ideas to ensure that they could quickly produce the requisite research reports when needed. What could be measured—the production of a set number of reports—had a detrimental effect on what was much more important to the firm, but could not be measured: the development of high-quality commercially relevant ideas. As the frustrated R&D manager at the firm noted, "The lack of objective criteria for monitoring the engineer's innovative work leads him to transfer his main attention from the quality of his ideas to attempts to find out whom to please and how."

Despite these concerns, over the past two decades, this historical reluctance to offer high-powered incentives to researchers appears to have ebbed considerably. The compensation for corporate R&D heads with titles like "chief scientific officer" changed dramatically in the 1990s, with much greater use of long-term incentives (e.g., restricted stock and stock options).[30] The ratio of the value of long-term incentives to total compensation for corporate R&D heads increased sharply over the period, from 25 percent of total compensation in 1989 to 51 percent in 1998. If we look instead at the (inflation-adjusted) dollar value of the long-term incentives offered, the amount almost tripled.[31]

Nor does this change appear to be confined to the top of the organizations. For instance, the ipPerformance Group recently undertook a survey of reward schemes for line scientists and engineers in 136 firms.[32] They found that, by far, the respondents saw monetary rewards as the most effective way to motivate employees—choosing these by more than three to one over nonmonetary recognitions. This attitude is very different from that of R&D managers in earlier decades.

The most dramatic manifestation of the increased interest in monetary rewards for innovation, however, has undoubtedly been the proliferation of contests. While some of these, such as the "Collaboration Across Cisco" award, have been internally focused, the most visible of these efforts have sought to garner ideas from outside the corporations. Some firms, such as IBM (Innovation Jam) and Procter & Gamble (Connect + Develop Organization), have turned directly to outsiders for help in solving complicated business problems, while start-ups such as InnoCentive and Genius Rocket seek to serve as intermediaries for others' R&D projects.

Taking Stock of Increased Incentives

But do we want more incentives? A standard economic view, rooted in the notion of "agency theory," suggests that stronger incentives would elicit more effort, and as a result more innovation. Researchers, as agents of corporate shareholders, have a temptation to slack off because they do not receive all the fruits of their labor. A bigger share of the pie for the scientists and engineers should induce them to work harder and smarter, and create more innovations.

These rewards might be financial or nonpecuniary, in the form of respect and recognition. But whatever the form, with more rewards should come better performance. In the context of R&D, pay linked to firm performance motivates researchers to work harder and to resist decisions that increase private benefits at the expense of shareholders, such as funding of "pet projects" or showing favoritism to select labs. This idea is not unique to economists: the organizational behavior literature has formulated a wide variety of "expectancy models," which hypothesize similar relationships between incentives, effort, and performance.

A variety of studies that look at more general settings show that incentive pay does appear to enhance performance. For instance, Casey Ichniowski, Kathy Shaw, and Giovanna Prennushi looked closely at forty-five steel finishing lines.[33] They concluded that the most productive lines were those where there was a strong element of pay for performance: where the line workers either received a stipulated share of the plant's profits, or where pay varied with the quantity and quality of output. The magnitude of this effect was substantial: switching from a traditionalist approach to one featuring incentive compensation and other progressive features could boost the profit per facility by $1 million annually, or even more. Similar results regarding the power of incentives have been found for workers ranging from auto-glass replacers to navy recruiters. Even professional golfers seem to respond to the size of the prize money pool, though this effect is seen primarily on the fourth and final day of championships, when presumably fatigue is setting in.[34]

But before we declare victory and pronounce incentives for researchers a good thing, we must consider the fact that innovation is a fundamentally different activity from replacing broken windshields or cajoling high school students into enlisting. In particular, to be successful, innovation requires an intensive, hard-to-monitor effort from the participants, often working together. Many researchers in social psychology have expressed concern that too-strong financial incentives may actually dampen interest in the activities being encouraged, which they term "motivation crowding-out." To use one oft-cited example, individuals may be more willing to donate blood when their duty to help their fellow man is invoked, rather than when these pleas are combined with the promise of a cash payment.

This point has been documented in many studies, from Harry Harlow and Edward Deci's experiments with problem-solving monkeys in the late 1940s to Sam Glucksberg's tests of human problem solving in the early 1960s. The foremost contemporary proponent of this view when it comes to innovation and creativity is my HBS colleague Teresa Amabile, a strong advocate for what she terms the "Intrinsic Motivation Principle of Creativity": that individuals will be the most creative when motivated by the challenge of the problem itself and not by external pressures. Just as artificial deadlines and a dismissive approach by managers to new ideas will kill off this intrinsic motivation, she argues, so too will an excessive emphasis on monetary rewards. In a number of articles, she and coauthors have found support for this view, such as one where she interviewed scientists about the circumstances around their most and least creative insights.[35]

Another critique of offering incentive pay for researchers stems from the importance of teams in these projects. In a setting where multiple researchers are performing complex, poorly defined tasks, apportioning credit may be virtually impossible. In this setting, scientists and engineers—who might anticipate that they will not be appropriately rewarded—will be reluctant to go all out in helping their colleagues. The difficulty of really determining who made these contributions may undermine the effectiveness of high-powered incentives. This argument—while the basis of a number of theoretical models— has also been articulated by many traditionalist R&D managers; as one related, "The company [does]n't want its engineers vying to be first, or at least be first to take credit. What would this do to teamwork?"[36]

Gustavo Manso has argued in a theoretical work that the truth lies somewhere in the middle, between these extreme views.[37] If incentives put too much emphasis on short-term success, researchers—and the managers who oversee them—will be excessively cautious. They will be unwilling to undertake risky projects that are likely to fail, but could yield very profitable insights if successful. Yet at the same time, pursuing innovation is a tough and time-consuming activity, which is hard for bosses to monitor. These individuals should receive long-run rewards if successful in their endeavors. He advocates a combination of short-term protections and long-term rewards: for instance, R&D managers might ideally receive "golden parachutes" that assure them of a substantial payout if fired but also sizable numbers of stock options with long vesting periods.

But how do incentives actually work in the real world? The evidence is not as conclusive as we would like. In large part, this lack of a definitive answer stems from the same problem: our inability to tell research directors what to do! Because we can only observe the outcomes and choices made in the real world—choices that may have been made as a result of all sorts of information we are not privy to—our ability to reach definitive conclusions about the impact of high-powered incentives on innovation is limited.

One body of work has looked at how monetary rewards interact with intrinsic motivation. It is hard to think of an arena where intrinsic motivation for innovation should more completely rule than open source programming. Many contributors to open source projects are self-styled rebels, as the following quotation from Richard Stallman, the creator of the General Public License that governs most open source projects, suggests: "The idea that the proprietary software social system—the system that says you are not allowed to share or change software—is unsocial, that it is unethical, that it is simply wrong, may come as a surprise to some readers. But what else could we say about a system based on dividing the public and keeping users helpless?"[38]

Many members—and observers—of the open source community have emphasized the importance of the nonpecuniary benefits as a driver of participation in these communities. For instance, computer programmer and open source advocate Eric Raymond argues: "The 'utility function' Linux hackers are maximizing is not classically economic, but is the intangible of their own ego satisfaction and reputation among other hackers . . . Voluntary cultures that work this way are actually not uncommon; one other in which I have long participated is science fiction fandom, which unlike hackerdom explicitly recognizes 'egoboo' (the enhancement of one's reputation among other fans)."[39]

Even in this setting, however, economists have documented the importance of explicit, if deferred, financial rewards.[40] To be sure, the traditional programmer working nights and weekends on an open source software development does not receive any direct pay. But eventually, open source contributions may lead to very real financial benefits, whether future job offers or shares in commercial open source–based companies.

In point of fact, the open source world—for all the rhetoric to the contrary—actually does seem to be a particularly good place to signal a high level of competence to the private sector. In an open source project, because each contribution is signed by the programmer who makes it, outsiders can see the contribution of each individual, whether that component "worked," whether the task was hard, if the problem was addressed in a clever way, whether the code can be useful for other programming tasks in the future, and so forth. (Contrast this with Apple, where one of Steve Jobs's first acts upon returning as CEO was to remove the credits to individual programmers in the firm's software, on the grounds that such a move would make it harder for outsiders to poach their best programmers.[41])

Studies of individual contributors suggest that successful contributors to open source projects—in addition to whatever intrinsic benefits they enjoy—also benefit financially. Jeffrey Roberts, Il-Horn Hann, and Sandra Slaughter, in an interesting study, examine the career paths of contributors to the Apache project, drawing on a wide variety of records of the project itself.[42] The authors explain the earnings of the contributors over time, controlling for information on the respondents' background, work experience, and contributions and position to the Apache project. They look at how compensation at his or her "real" job changes as the individual climbs the ladder at the open source project. The results suggest that the sheer volume of contributions to the Apache project have little impact on salary. But individuals who attain high rank in the Apache organization enjoy substantially higher wages, whether or not their work directly involves the Apache program. The results suggest that even in a setting where intrinsic considerations would be expected to dominate, financial rewards appear to be a powerful spur to innovate.[43]

Similarly, in laboratory experiments, it has been shown that creative exploration can coexist with incentive compensation—as long as the incentives are designed appropriately. One fascinating illustration of this point is in a laboratory experiment run by Florian Ederer and Gustavo Manso.[44] In this study, the students were asked to run electronic lemonade stands over twenty periods. There were three different groups: one that received a fixed payment regardless of the sales of lemonade, one where the payment was linked to the outcome in each of the twenty periods, and one where the compensation was based only on the profits produced in the last ten periods. The secret of the game was that the initial location of the stand, while reasonably profitable, was not the ideal one. Subjects needed to move dramatically away from the default selling location to find the prime spot, and hence earn much more from their lemonade venture. Those who had long-term compensation contracts—where the pay was dependent only on where the stand ultimately ended up, not the steps along the way—were far more likely to explore adventurously, and as a result to find the right location, than either those receiving a fixed payment or a payment linked to the profits in every period.

An illustration of the powers of long-term incentives in the real world is found in another analysis by Manso, this time with Pierre Azoulay and Joshua Graff Zivin.[45] The authors contrast two ways that elite biomedical scientists get funded in the United States. Most common are grants awarded by the National Institutes of Health. These grants are relatively short term (typically three years) and highly competitive to win and renew. Alternatively, brilliant scientists can be appointed as investigators of the Howard Hughes Medical Institute (HHMI). Although they continue to work at the same university, HHMI researchers face a very different set of incentives. They are explicitly told to "change their fields" and are given the resources, time (five-year renewable appointments, with a lax first review and a two-year phase-down in case of termination), and autonomy to accomplish this. Compared with a set of equally eminent scientists with similar resources, the HHMI program appears effective in boosting the rate of discoveries, particularly of the highest-impact scientific papers. The impacts of the HHMI awards are even larger for other outcomes, such as the grooming of the next generation of cutting-edge researchers. While ultimately, the HHMI-backed scholars have strong incentives—unless their research labs deliver, the grants will be terminated—the longer-term nature of the schemes allows them to choose more productive routes.

Finally, Julie Wulf and I have documented the relationship between the compensation of the top corporate R&D chieftain and innovation at that firm.[46] Among the standard firms (with a centralized R&D lab), a clear relationship emerges: more long-term incentives granted to corporate R&D heads are associated with more heavily cited patents. These incentives also appear to be associated with more frequent awards and more fundamental patents. Meanwhile, consistent with the arguments above, we find little association between short-term incentives for corporate R&D heads and innovation.

Again, we must be cautious in interpretation of any individual work, but taken together, the studies suggest that appropriately designed long-term incentives can boost innovation.

Final Thoughts

Viewed with the benefit of hindsight, and particularly in light of recent research into the organizational settings that are most conducive to innovation, the changes to corporate R&D that have unfolded over the past two decades do not seem particularly grim. To be sure, there are many issues and uncertainties surrounding how to create an environment conducive to innovation. But the preponderance of the evidence suggests that the claims of imminent harm have been overstated.

The next section switches focus, turning to consider the primary alternative model to funding innovation that has emerged over the past few decades, the venture capital approach.

Part Two

The Venture Alternative

4

Getting Venturesome

I F ONLY MOST companies were like Facebook—or at least the version we see in the movie *The Social Network*. Then they would find helpful venture capitalists competing to give them dollars, along with knowledgeable angel investors like Peter Thiel. But alas, for almost all high-growth firms, financing is a challenge even when things go well.

Consider, for instance, Suniva, a solar company spun out from Georgia Tech in 2006.[1] Despite having a novel single-crystal silicon solar cell technology, an exclusive license to almost forty patents and applications from the university, and the leadership of a world-class engineering professor, raising financing has been an ongoing challenge for the firm. This reflects the fact that, throughout its existence, considerable uncertainty has surrounded Suniva's business model and the speed with which it will generate profits. Although the firm has succeeded in advancing its technology and in raising $130 million in venture capital from high-caliber groups such as New Enterprise Associates and Warburg Pincus, as of 2011, the process has not been easy for the firm or the investors. Key strategic decisions—such as whether to accept a loan from the Department of Energy, which would have required it to build its manufacturing facility in the United States—have required careful study and discussion.

And so it goes for most firms that seek funding through the venture capital (VC) model. By VC, I mean a group of individuals making equity or equity-like investments in growing firms, using funds raised from other people or institutions. The individuals running these firms have had to grapple with many of the same issues as the corporate research managers: highly uncertain projects run by knowledgeable, enthusiastic, and often overoptimistic champions. If things work out, there may be riches for all, but if not, all the sponsors are typically left with is some difficult-to-sell intellectual property. Figuring out which projects to allocate resources to and which to walk away from has been daunting in both corporate and venture models. But the approaches taken by venture firms, as we will explore in depth in this chapter, have been very different from those of their corporate brethren.

To get our arms around the history of the venture capital model—which, arguably, has generated some of the greatest wealth the world has even seen—the rest of this chapter condenses its long and complex history into a few pages. To keep the length manageable, I focus on four key milestones, from venture capitalists' first appearance (or nearly so), to the creation of the fund model, the industry's global spread, and, ironically, the recent choice of many funds to remain small.

Milestone 1: Creating the Investment Model

Like many great successes, venture capital has many putative inventors. J. H. Whitney, established immediately after World War II, has described itself as the first venture firm. So has Bessemer Venture Partners, which began investing the Phipps' family money in 1911. Others have traced the origin of venture investments to the tenth-century *commenda* contracts of the Genoese and Venetian trading families, or even further back to the Byzantine *cheokoinonia* or the Muslim *muqarada*. The most ambitious have pointed back four millennia to the Code of Hammurabi, which laid out the essential ground rules for these activities.[2]

Rather than adjudicating this particular dispute, we will focus on the most widely accepted choice, American Research and Development (ARD).[3] The genesis of this fund dates to 1945, when a committee of leading Boston citizens came together to consider the challenges of the region and the nation in the postwar era. In their deliberations, the committee was largely motivated by fears about innovation and economic growth. Prior to World War II, the bulk of the financing for entrepreneurial ventures—which were typically too risky for banks to lend to or to go public on a stock exchange—came from wealthy individuals. But these individuals were frequently poor investors, being distracted by their far-flung business interests and wide-ranging other pursuits. The economy needed new firms, the committee felt, to commercialize the promising new technologies developed during the war. But without some additional avenues to raise financing, they feared, these new entities would not get off the ground. The failure to raise capital would depress the economic health of the region and the nation, and might even lead to a relapse to the debilitating conditions seen in the 1930s. Although proposals for aggressive government intervention to fund entrepreneurs had been floated in Washington, the committee objected to these ideas on both practical and philosophical grounds, instead proposing a business to finance new businesses.

Within a year, Georges Doriot had assumed the leadership of ARD. Much has been written about Doriot's many accomplishments: an influential Harvard Business School professor, a brigadier general who played a key role in designing the supply chain for U.S. troops in World War II, a cofounder of INSEAD, the leading European business school, and the like, but surely laying out the template for modern venture capital investing was one of his greatest feats. From its earliest communications, ARD emphasized three aspects of its role: sorting, oversight, and certification.

SORTING. Unlike a bank, which might fund one-half or one-third of the firms that walked in the door, ARD set the bar extremely high. Typically, it funded only 1 to 2 percent of the business plans it received. Even then, it typically made small investments, with additional funds conditional on the firm's making satisfactory progress in advancing its ideas.

OVERSIGHT. Unlike a banker, who would be satisfied with quarterly statements, Doriot emphasized intensive involvement in the companies in which ARD invested. As the firm noted in its promotional material, "The Corporation usually will be represented on the Board of Directors, will assist in the procurement of personnel for key positions, and will endeavor to make available the best possible . . . assistance."[4] One aspect of this tight governance was that investments were often structured as a preferred debt note, in addition to common stock, which gave the venture capitalists more power—and put more pressure on the entrepreneurs—than had ARD instead just held common stock.

CERTIFICATION. Doriot realized that the path to success for new ventures was not easy. Frequently, these firms lacked the track record or the connections to get attention from potential strategic partners, investment banks, and other actors. By pairing these firms with seasoned mentors who had both credibility and networks, he hoped to give the firms "stamps of approval" that could enhance their probability of success. As he stated, "A team made up of the younger generation, with courage and inventiveness, together with older men of wisdom and experience, should bring success."[5]

The prescience of Doriot's vision has been borne out in the ensuing decades, as contemporary VCs have essentially followed his lead in financing and guiding firms. Economists Steve Kaplan and Per Stromberg have examined how venture capitalists think about prospective investments, as well as the dance between investors and entrepreneurs. Drawing from the memoranda that venture funds use to evaluate individual deals, they show that while external factors (such as the evolution of the market) are certainly important, typically the most critical considerations driving whether or not to invest come back to the personal considerations that Doriot highlighted: the depth of the entrepreneur's knowledge, his or her behavior, willingness to work hard, and ethics.[6]

Meanwhile, the VC's amount of control varies. Sometimes VC funding comes with seemingly onerous provisions regarding the ownership of the equity, control of the board, and division of the proceeds if the firm is liquidated.[7] But these features are not simply complex and onerous for their own stake. Kaplan and Stromberg show how, consistent with Doriot's vision, the features of these agreements address the challenges inherent in financing and overseeing young, high-risk entities: many of the prerogatives the venture capitalists enjoy depend on the firm's performance. If the firm does poorly, their venture investors' rights to seats on the board and votes on key decisions will grow to the point where they have full control. (The investors' share of the proceeds in the event the firm struggles or fails will also frequently increase.) As the firm does better, more and more of the control reverts to the entrepreneur. If the firm does very well, and goes public with an attractive valuation, the venture investors will retain their equity stake, but all their other rights will disappear.

The key motivating principle, Kaplan and Stromberg argue, is that venture capitalists need to take control only when things are uncertain or problematic, and be willing to relinquish it in other circumstances. Swings in control are even more dramatic in cases where uncertainty might be particularly high, such as early-stage firms.

One of the most common control mechanisms VCs use is meting out financing in discrete stages over time, during which the prospects for the firm are periodically reevaluated. Staged capital infusion keeps the owner/manager on a tight leash and reduces potential losses from bad decisions. The amount of time between individual rounds of financing varies with the riskiness of the venture and the investors' need to gather information. Paul Gompers, for instance, shows that early-stage firms receive significantly less money per round.[8] Ventures with more hard assets like factories and equipment—where presumably it is easier to determine what the firm is doing with the funds—receive larger and less frequent financing rounds, while those where the main value is just ideas get small rounds. These results suggest the important monitoring and information-generating roles played by venture capitalists, who want to make sure their investment is safe, again along the lines that Doriot articulated.

The advice and support venture capitalists provide is often embodied in their role on the firm's board of directors. Over half the firms in one sample have a venture director with an office within sixty miles of their headquarters, suggesting the hands-on monitoring that these investors provide.[9] Again, there's a clear move toward more control as the entrepreneurial venture gets riskier. In times of crisis—when a top manager at an entrepreneurial firm is replaced—venture capitalists are far more likely to be added to the board than when things are going well, a pattern not seen among the other directors. In short, Doriot's vision of hands-on engaged venture investors is very much alive today.

Milestone 2: Creating the Fund Model

General William Draper, a long-serving military officer and occasional industrialist, founded Draper, Gaither & Anderson (DGA) in 1958 along with Rand Corporation founder Rowan Gaither and air force general Frederick Anderson.[10] DGA was notable in part for its location (it was one of the pioneering firms in Silicon Valley), but particularly for its structure. With a small number of private backers—of its $6 million fund, the venerable investment bank Lazard Frères contributed $1.5 million and the Rockefellers and two other leading families nearly as much—it eschewed the publicly traded structure of earlier funds. Rather, it adopted the model that would become standard in the industry, the limited partnership.

In making this choice, the partners were swayed by the difficulties that ARD had faced: for much of its first fifteen years, Doriot had struggled to raise capital. After the $500,000 ARD had raised from its initial backers, additional funding had been slow to come by. Its profile as a non-dividend-paying entity with losses stretching out for the foreseeable future did not excite many institutions. Moreover, the frequent articles about ARD's social goals of promoting economic development and entrepreneurship raised concerns among investors that financial returns were second order. Early on, a decision was made to raise money by taking ARD public. Even then, no investment bank was willing to underwrite the offering, and it was largely sold off in small blocks to retail investors on a "best efforts" basis. Over the course of the 1950s, the firm frequently dealt with disillusioned investors in the face of market turbulence and portfolio companies that took longer than expected to mature.

While much of this dissatisfaction may have reflected the naiveté of inexperienced investors, even knowledgeable investors criticized ARD's tendency to stick with troubled firms too long. Doriot defended himself spiritedly: "We are really doctors of childhood diseases here. When bankers or brokers tell me I should sell an ailing company I ask them, 'Would you sell a child running a temperature of 104?' "[11] But the experience of "zombie" firms like Ionics raised the question of whether the evergreen structure that characterized the ARD fund was appropriate: once the capital was raised, ARD had no obligation to ever return the funds, and could keep on investing until the firm ran out of money or was acquired.

The limited partnership structure that DGA introduced to the venture capital industry addressed several concerns. By raising funds up front, the venture capitalist received all the money he or she would need for the life of the fund. There would be no need to go "back to the well" every year or two for another tranche of capital, as Doriot had been required to. (The one exception related to the fact that investors in venture capital limited partnerships are not typically required to put in all the capital that they commit to the fund up front. Rather, they promise the funds, and then the general partners draw down the capital as needed. In extraordinary periods, such as after the financial crisis of 2008, there may be real worry that the investors cannot or will not meet their commitments.)

Moreover, the new firm's founders felt that the ARD structure had brought about excessively conservative thinking. As Pete Bancroft, one of the junior investment professionals at DGA noted of their predecessors, "They did not dare as greatly or as well."[12] Much of this could be attributed to the compensation scheme at the Boston firm: everyone, including Doriot, received a salary—there was no explicit link of compensation to performance. Instead, at DGA, not only were the partners major investors (contributing $700,000 of the capital, or more than 10 percent of the funds), but they also got a significant profit share. In particular, the DGA partners received 40 percent of the capital gains (well above the 20 to 30 percent standard in the industry today), in addition to their proportionate share of the amount going to the limited partners.

But at the same time, the new structure held advantages for the limited partners. The venture capitalists raised funds only for a set period, typically seven to ten years, with the possibility of an extension for a few more. This stipulation provided a distinct time limit for the venture capitalists' activity: there would be no nurturing sick firms indefinitely. Moreover, the investors had the assurance that, ultimately, they would get back whatever money remained. While it would be very hard to dislodge the manager of an "evergreen" fund like ARD without an expensive battle for control of the board, limited partnerships simply wind up. Unless a venture firm can convince investors to ante up for a new fund, they will go out of business.

The partnership structure also addressed investors' concerns by drawing a sharp line between the limited and general partners. The limited partners were limited in the sense that their liability was capped by the amount they invested: if the fund invested in a biotechnology company whose drug unfortunately ended up killing several people during a clinical trial, the general partners running the fund would face many millions of dollars of claims for damages, but the investors would not. This was important to many investors, as someone putting a few million dollars into a venture fund does not want to have to worry about the risk of losing millions more if something goes wrong with a high-risk investment.

At the same time, the fact that the general partner needed to return to the limited partner for funds meant that the institutional and individual investors who provided the capital could have a lot of power. Bill Draper, General Draper's son and the eventual founder of Sutter Hill and a number of other venture groups, related how, while working for his father at DGA, they were approached about investing in the first condominium development in Hawaii.[13] The investment appeared to be highly promising, and proved to be very successful. But midway through the due diligence process, Draper was summoned to Rockefeller Center and dressed down by one of the family office's partners for investing outside of their promised mandate and expertise. He was told in no uncertain terms that the Rockefellers could access such investments through much more knowledgeable, real estate–focused intermediaries. As a result, Draper was forced to turn down the investment, which ultimately provided a higher return than any of the deals that ended up in DGA's portfolio.

This give-and-take between limited and general partners in venture firms has been familiar territory since the industry's early days. In part, the features in contracts between venture groups and their investors are driven by the need for oversight: less-established groups, where there is more uncertainty about the ability of the investment team, tend to grant the general partner less control, such as terms that limit the kinds of deals the fund can invest in or other activities that the venture capitalists can engage in.[14]

But the numbers of venture capitalists and investors are small, so the structure of these deals is also shaped by supply and demand. A sudden increase in demand for venture capital investing services—if, say, institutional investors suddenly increase their allocation to venture capital funds—should increase bargaining power in favor of the venture investors. Only a small number of venture capitalists may be raising a fund at any particular time, and these firms are likely to be differentiated by size, industry focus, location, and reputation. Meanwhile, managers who allocate alternative investments for institutions often operate under limitations about the types of funds in which they can invest (for instance, rules that prohibit investments into first funds raised by venture organizations) and are pressured to meet allocation targets by the end of the fiscal year. As a result, venture groups whose services are in more demand—whether because of a hot market or a successful track record—enter into agreements with more freedom and also with higher fees.

Undoubtedly, the creation of the limited partnership, which addressed many of the drawbacks of evergreen funds such as ARD, represented an important breakthrough. But limited partnerships also posed issues of their own: these partnerships too could be swayed by the whims of investment fashion.

Milestone 3: Going Global

The first two milestones focused on events in the United States, reflecting the reality of those years when American firms dominated the venture industry. Our third milestone—the creation of Advent International's global network of funds—was a key step toward breaking this mold.[15]

Advent was very much the brainchild of Peter Brooke, who had begun his career doing small business lending at the First National Bank of Boston. As a junior analyst, he had pushed the bank to shift its focus to the emerging high-technology firms in the Boston area. Not only did this move prove profitable for the bank, but it triggered Brooke's interest in the nascent venture capital industry. His first venture firm, TA Associates, had initially focused on making equity investments in these same firms, but by the early 1970s, Brooke was increasingly interested in extending the venture model overseas. While still running his U.S.-focused fund, he began meeting with European policy makers and investors about the possibility of establishing affiliate funds. In so doing, he was following in the footsteps of Georges Doriot, who had established a European affiliate fund. Unlike Doriot's pan-European effort, which proved to be ahead of its time along a variety of dimensions, Brooke focused on country funds, with well-connected investors and powerful institutional backers.

This process of building affiliates greatly accelerated in 1984, when Brooke spun off Advent International from TA Associates. Here, he refined a model that he termed the Advent Network. While most local venture capital groups outside the United States at the time suffered from limited resources and a somewhat parochial worldview, his network would provide several advantages. Advent would help its partners with due diligence, provide access to a network of executives and corporations worldwide, and give guidance through its seat on the funds' boards and investment committees. The U.S. fund would also greatly facilitate fund-raising, since a number of institutions that would otherwise be unwilling to invest in a faraway fund would do so once Advent was involved. Affiliate funds were established not just in Europe, but throughout Asia and Latin America.

One clear success was Advent's foray into Israel in the early 1990s. At the time, entrepreneurial activity there was booming, particularly in the software industry, but there was only one small and not particularly successful venture group active in the market. The Israeli government sought to spur venture capital activity there, launching the Yozma program to provide matching funds to foreign venture groups who were willing to set up local funds. The key government officials realized that Advent would be a perfect fit, due to its experience with overseas partnerships.

To turn this vision into reality, Advent teamed with the Discount Investment Corporation, a publicly traded holding company with experience undertaking active investment in Israeli entities (though typically not start-ups). They created a new venture fund, Gemini Israel, that had a complex two-fund structure, which allowed them to both support firms domestically and finance their overseas activities (the Israeli government funds could not be used to finance companies outside the country). In addition to helping recruit Israeli executives seasoned by working with entrepreneurial firms, Advent contributed two members of Gemini's board: Brooke and Clint Harris, an Advent cofounder responsible for the management of the initiative. From its modest $36 million first fund, Gemini has gone on to raise over $650 million in four subsequent funds.[16]

Advent, like the other groups that employed these models, faced some tough challenges as the model developed. One of the major trade-offs was the choice between far-flung affiliates and centralized activities. While the affiliate groups had strong local knowledge, the results were frequently heterogeneous, and the relationships with its partners became strained. Moreover, as Advent began raising larger funds to invest internationally, it felt a growing desire to have its own people on the ground in these markets. This change naturally introduced strains in the relationships with the affiliates, who gradually weakened their ties with the parent organization.

A second major turning point was the firm's abandonment of venture capital investing in many markets. In Europe and much of Asia, Advent realized it was garnering far better returns from its investments in mature firms, frequently structured as leveraged buyouts, than its seed and early-stage deals. Moreover, successive iterations of Advent's flagship global fund raised $315 million in 1993, $1 billion in 1997, and $1.5 billion in 2001. With this much capital to deploy, it became increasingly difficult to do early-stage transactions. Given that each portfolio company would require a considerable amount of management attention, regardless of size, the minimum size for each deal crept inexorably upward.

In this respect, the experience of Advent was representative. Venture capital (as opposed to private equity such as buyouts and expansion capital) still remains dominated by activity in the United States. Figures 4-1 and 4-2 show the distribution of venture capital investments in 2010, in absolute levels and as a share of GDP. The first chart makes clear that the United States still has the lion's share of the activity in this sector, with the rapidly growing Chinese market still a distant second. The activity in the United States represents more than twice that in all the other tabulated nations combined.

When we look at activity as a share of the economy in the second chart, the dominant position of the three nations is clear: Hong Kong, the United States, and Israel. The venture investment total in the first

FIGURE 4-1

VC investments (US$ bil.)

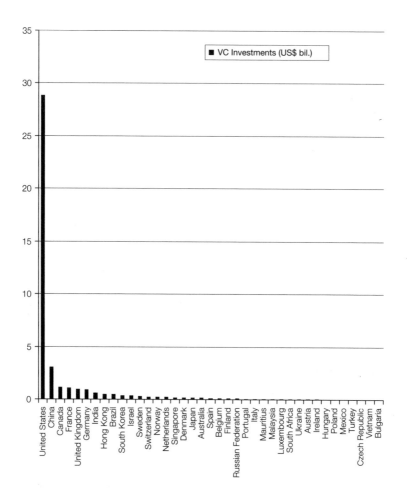

Source: This figure is compiled from various publications and websites of the Canadian, European, Indian, Israeli, and U.S. (National) venture capital associations, as well as Thomson Reuters, "VentureXpert Database," http://www.venturexpert.com. In some nations where venture capital investments are not clearly delineated, I employ seed and start-up investments.

nation must be regarded with caution, as much of the funds may be intended for enterprises based in China but domiciled in Hong Kong due to regulatory and tax considerations. (Two nations where the vast majority of investments have the bulk of their economic activity

FIGURE 4-2

VC investing as a percentage of GDP

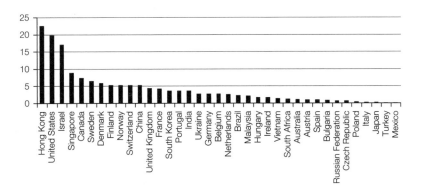

Source: This figure is compiled from various publications and websites of the Canadian, European, Indian, Israeli, and U.S. (National) venture capital associations, as well as Thomson Reuters, "VentureXpert Database," http://www.venturexpert.com. In some nations where venture capital investments are not clearly delineated, I employ seed and start-up investments.

elsewhere, Luxembourg and Mauritius, are excluded from the table.) The United States stands well ahead of the other major industrialized nations on this measure. These comparisons can be sensitive to exactly how venture capital is defined: for instance, whether investments in already profitable but rapidly growing firms are included. But clearly, despite the trend toward global venture investments, the industry still has a long way to go in terms of internationalization.

What lies behind the apparent failure of venture capital to take root in many markets? Some clues can be found in the differential success of venture capital groups across nations. A number of studies have sought to look carefully at which factors drive both the volume and the ultimate success of these investments.[17] These works highlight three key factors:

1. *The availability of capital.* Despite the increasingly global nature of capital flows, having a dedicated pool of investors locally is important: local funds often are more inclined to

invest in a market at times when international investors shy away. As we see from the Advent history, local institutions and individuals can bring important connections to the table. In many markets, the growth of venture activity has been retarded by the unwillingness of local institutions to consider these investments.

2. *Regulations that create a favorable environment.* As we will discuss in more detail in chapter 7, without the freedom to enter into enforceable contracts and adjust firms' labor forces, entrepreneurs are unlikely to succeed.

3. *Successfully getting out of investments.* Unless investors believe they will get their money back (and then some), fund-raising will be very hard. Countries with an active market for public offerings again seem to enjoy a leg up on their peers.

Although the transplanting of the venture capital market overseas was an important step, the transition is far from complete. In many markets, venture capital has found it hard to take root. And while there have been some promising developments in recent years—of which the gravity-defying growth of the Chinese market is notable—considerable ambiguities remain. Neither the extent to which venture capital will become a global industry nor the ideal structure of global venture funds is certain.

Milestone 4: Going Small

Traditionally, venture groups were different sizes. A seed fund would do the first institutional investment round, joined later by a balanced fund, and perhaps by a later-stage specialist just before the firm went public. Naturally, the seed funds were much smaller than later-stage ones, since the financing rounds where they played a key role were so much smaller. But in most cases, the smaller groups were not small because they wanted to be. They did seed investing because they didn't have the resources to do other types of transactions. These small groups typically aspired to raise progressively larger amounts of funds, and gradually moved away from seed investing. The main reason that many early-stage investors wanted to get away from this space was that returns were typically quite poor. Groups would undertake the hard work of identifying promising seedling firms and purchase an equity stake in a first round of financing, but then see their ownership of the company substantially watered down when they did not have the funding to take part in later financing rounds.

The past few years have seen a dramatic bifurcation in the venture industry. Some groups have become aggressively larger, such as Bessemer and New Enterprise Associates, which recently raised funds well north of $1 billion. Other venture organizations raised a number of separate funds that collectively represent a significant amount of resources. Sequoia Capital between 2007 and 2010 raised seven separate funds—geared to both early- and late-stage investments and targeting specific geographies such as China, India, and Israel—that totaled over $4.25 billion.[18]

At the same time, a number of groups—"superangels" or "micro VCs"—have deliberately focused on smaller funds and deals. These have argued that their small size gives them a kind of flexibility and focus that their larger brethren do not have. For instance, Mike Maples raised only $84 million for his second Floodgate Fund despite having terrific success with his first fund (among its investments were Twitter, the textbook rental firm Chegg, and iPhone game developer Ngmoco, which was acquired in 2010 for $400 million). He explained his philosophy:

> If you want to know a fund's strategy, you have to ask one and only one question: How big is your fund? . . . What ends up happening

is that the size of the fund dictates all the decision-making. People say you can do super-early and super-late stage investing at the same time, but the fact of the matter is that you spend most of your attention where the money goes. So the constraint on the fund size is part of the discipline of being purely early stage. We are not a small fund so that we can become a big fund.[19]

Probably the most dramatic illustration of this trend has been Y Combinator, a start-up incubator "applying mass production techniques to venture funding" that was founded in 2005.[20] It invests relatively small amounts of money (around $18,000 on average) in a large number of start-ups (fifty to sixty per year) for minority equity stakes (6 to 7 percent on average). The company is run by four full-time and two part-time partners, who collectively work with the entrepreneurs in an intense graduate school–like atmosphere during a three-month period to help refine their products and open doors to more fundraising. The biggest asset of Y Combinator that draws in applicants is not the financial support; rather, it is the partners' wealth of practical knowledge and connections.

The partners at Y Combinator work with the founders during their months together. Operating under the "launch fast and iterate" philosophy, they offer an unlimited amount of "office hours" to help steer the direction of the start-ups. The partners individually work with founders to elucidate company vision through a "whiteboard pitch" that usually becomes the basis of their presentations to investors.

Participants in Y Combinator also offer feedback on each other's ideas throughout the boot camp. Cofounder Paul Graham emphasizes the value of peer evaluation, stating, "[T]here is a certain kind of brutal candor only peers can deliver, and if there's one topic where it's warranted, this is it."[21] First, during "Prototype Day," which takes place four weeks into the program, the founder(s) of each company introduce their idea to all other founders and a vote is conducted to determine which start-up is the most attractive from an investor point of view. Next, during "Rehearsal Day," which takes place a week before "Demo Day," the founders present a more refined version of their pitches and another founder-deemed "winner" is declared. The entrepreneurs also present in front of top alumni ("Alumni Day") and in front of Sequoia Capital partners to receive advice and polish presentations.

Fund-raising is arguably the foremost advantage selected applicants gain from entering the program, as a considerable number of investors have sought, formally or informally, to target Y Combinator graduates. In the summer 2010 batch of start-ups, for instance, thirty-three of the thirty-five start-ups that needed to raise money post-YC did so. This process is encouraged through a variety of interactions. During each class, an "Angel Day" is held so that the founders can informally practice their investor pitches. Founders are matched with two angel investors and often stay in communication with them for the coming months. The Y Combinator program culminates with a formal presentation of the founders' ideas to hundreds of investors on "Demo Day." Throughout the process, the founders share their input on which investors individual founders should pursue, given their prior experiences; attempt to translate sometimes ambiguous messages by investors; and often talk to investors directly to encourage a deal.

Graham argues that the YC alumni network "is probably the most powerful network in the start-up world . . . not just because of its size, but also because its members have such a strong commitment to helping one another."[22] The success of the many alumni is also helpful. In addition, the partners have connections with various companies that are willing to provide complimentary services for Y Combinator companies and arrange an entrepreneur speaker series.

Y Combinator companies are set up (incorporation, vesting, intellectual property, etc.) by professionals to ensure that all the start-ups can grow on a legal foundation. In addition to freeing up entrepreneurs to work on their companies rather than on paperwork and helping avoid legal issues caused by founder ignorance, this step allows investors to conduct less due diligence. As of the summer of 2011, Y Combinator's top twenty-one companies were collectively valued at $4.7 billion.

Many of the rationales for micro venture capital funds lie in the changing economics of start-up firms. Advocates argue that today it costs far less to undertake a "lean start-up" than ever before, due to the availability of open source software, offshore programmers, and the ability to rent servers and other infrastructure on a "pay-as-you-need-it" basis. These claims appear reasonable, but even if we look at an earlier era before such innovations, there is evidence that supports the opportunity for such funds.

To be sure, there is persistence in venture capital—the best managers outperform again and again, a pattern documented by Steve Kaplan and Antoinette Schoar, who examined the extent to which strong performance in a sponsor's private equity fund predicted superior returns in subsequent funds.[23] Looking at 746 funds raised between 1980 and 1994, they saw a pattern of strong persistence: a fund that outperformed its peers by 1 percent annually could be expected to beat its rivals in its next funds by somewhere between 0.5 and 0.7 percent per year. The laggards, on average, also continue to perform poorly.

These results might lead us to conclude that the best approach would be to just stick with the long-established winners, rather than trying our luck with new and unproven groups. But this view is too simple. In many cases, successful firms tend to increase dramatically in size. And this growth in capital under management is often problematic.[24] Growth has a substantial negative effect on returns: a doubling of fund size, all else being equal, leads to a drop in IRR of over 5 percent.[25] Consider a group that managed a $100 million fund and ended up with a very attractive annual rate of return of 25 percent. If the partners then went out and raised $200 million for the next fund, on average we would expect the fund to perform considerably worse, with a yearly return of under 20 percent.

While new funds by seasoned, already successful groups can be expected to outperform, in many cases this success leads groups to sharply increase capital under management, often precipitating a deterioration of returns. Because many of the firms that have grown rapidly have been successful in the past, this pattern limits the persistence that Kaplan and Schoar documented. Were it not for this tendency, they would likely find an even stronger relationship between past and future performance. And it suggests that beyond the arguments about a new model for "lean start-ups," there is a strong case for groups that can stay small and focused and resist the temptation to grow.

The Bottom Line

For most of the past three decades, investments made by the entire venture capital sector totaled less than the R&D and capital expenditure budgets of individual large companies such as IBM, General Motors, or Merck. So it would be reasonable to be skeptical of whether the VC model represents an alternative model of organizing innovation.

One way to explore this question is to examine the impact of venture investing on wealth, jobs, and other financial measures across a variety of industries. Even if we just look at the subset of firms that went public and remain traded today (consistent information on venture-backed firms that were acquired or went out of business is hard to find), these firms have had an unmistakable effect on the U.S. economy. In late 2011, 677 firms were publicly traded on U.S. markets after receiving their private financing from venture capitalists. These firms made up over 11 percent of the total number of public firms in the United States at that time.[26] And of the total market value of public firms ($25.9 trillion), venture-backed companies came in at $2.3 trillion—9 percent. Venture-funded firms also made up 4 percent (over $1 trillion) of total sales ($24.7 trillion) of all U.S. public firms at the time. Finally, those public firms supported by venture funding employed 6 percent of the total public-company workforce—most of these jobs high-salaried, skilled positions in the technology sector. Clearly, venture investing fuels a substantial portion of the U.S. economy.

Venture investing also strengthens particular industries. To be sure, it has relatively little impact on industries dominated by mature companies—such as the manufacturing industries. That's because venture investors' mission is to capitalize on revolutionary changes in an industry, and the mature sectors often have a relatively low propensity for radical innovation. But contrast those industries with highly innovative ones, and the picture looks completely different. For example, companies in the computer software and hardware industry that received venture backing during their gestation as private firms represent more than 75 percent of the software industry's value. Venture-financed firms also play a central role in the biotechnology, computer services, and semiconductor industries. All of these industries have experienced tremendous innovation and upheaval in recent years.

The impact on innovation seems particularly substantial. Thomas Hellmann and Manju Puri show this by comparing 170 recently formed firms in Silicon Valley, including both venture-backed and nonventure firms.[27] Using questionnaire responses, they find evidence that venture capital financing is related to product market strategies and outcomes of start-ups. Firms that are pursuing what they term an "innovator strategy" are much more likely to obtain venture capital, and to get it faster. The presence of a venture capitalist is also associated with significantly less time to bring a product to market, especially for innovators (probably because these firms can focus more on innovating and less on raising money). Furthermore, firms are more likely to list obtaining venture capital as a significant milestone in the life cycle of the company as compared to other financing events. The results suggest significant interrelations between investor type and product market dimensions, and a role of venture capital in encouraging innovative companies.

In my work with Sam Kortum, I visit the same question in a different way, looking at whether the participation of venture capitalists in any given industry over the past few decades led to more or less innovation.[28] Venture funding does have a strong positive impact on innovation. On average, a dollar of venture capital appears to be *three to four* times more potent in stimulating patenting than a dollar of traditional corporate R&D. The estimates therefore suggest that venture capital, even though it averaged less than 3 percent of corporate R&D in the United States from 1983 to 1992, is responsible for a much greater share—perhaps 10 percent—of U.S. industrial innovations in this decade. In short, it seems that the venture model does appear to have a substantial economic development impact.

Final Thoughts

The bottom line is that intermediaries do seem to play a powerful role in boosting innovation in the firms they fund and oversee. At the same time, as a model of funding innovative research, venture capital has some important limitations. We have already highlighted one of these: the limited success that these funds have had in most nations outside the United States. The next chapter will dig deeper into the limitations of the venture model, and highlight that although it may be a powerful way to fund and oversee innovation, venture capital is not a universal solution.

5

The Shortcomings of Venture Capital

THE ABUNDANT PROPAGANDA—no, I mean white papers—produced by the National Venture Capital Association, the industry's lobbying body, concludes that the path to more innovation is simply to have more venture capital. As their 2011 *Venture Impact* report intones, "Venture has proven itself to be the most effective mechanism for rapidly deploying capital to the most promising emerging technologies and industries."[1] What could be simpler than prescribing more of the same?

But sadly, venture capital is far from a cure-all when it comes to innovation. Despite its many virtues, it has four essential constraints—its scope, the seemingly inevitable boom-bust cycle, mercurial public markets, and uneasy oversight—that limit its ability to promote innovation.

A Limited Scope

Over the years, venture investors have financed a progressively narrower range of technologies. Examining the portfolio of a single venture group three decades apart demonstrates the extent of the narrowing.[2]

Charles River Ventures was founded by three seasoned executives from the operating and investment worlds in 1970, in collaboration with MIT and the West Coast–based Mayfield Fund. The team described themselves as "creative builders" who looked for growth companies that would generate at least a 25 percent rate of return.

Within its first four years, it had almost completely invested its nearly $6 million first fund in eighteen firms. The ultimate outcomes of these investments were not that unusual: the portfolio included some companies that went public, others that were acquired for a range of valuations, and some that the venture capitalists wrote off. What was striking, however, was the range of industries covered.

To be sure, these included classes of technologies that would be comfortably at home in a typical venture capitalist's portfolio today: a start-up designing computer systems for hospitals (Health Data Corporation), a software company developing automated credit scoring systems (American Management Systems), and a firm seeking to develop an electric car (Electromotion, which, alas, proved to be a few decades before its time). Other companies, however, were much more unusual by today's venture standards: for instance, start-ups seeking to provide birth control for dogs (Agrophysics), high-strength fabrics for balloons and other demanding applications (N. F. Doweave), and turnkey systems for pig farming (International Farm Systems). In total, only eight of the initial portfolio companies—or less than half—related to communications, information technology, or human health care.

The fund's portfolio looks very different in 2012. As before, it contains some firms whose success is all but assured, such as Twitter. Other private firms, such as iControl, whose software allows utilities, broadband providers, and home security companies to offer interactive services to residents, appear to have very promising prospects. The success of others is less certain.

But when we look at the industry mix of the portfolio, dramatic differences from 1974 are apparent. Of the firms listed as investments in
July 2011, fully 77 percent were in three categories: software, social
networks/new media, and communications. Of the remainder, five
were in the energy space (mostly cleantech), and the remainder in a
variety of fields: two each in industrial products, semiconductors, and
intellectual property licensing, and one in finance.

What has happened to Charles River's portfolio reflects what has
taken place in the industry at large.[3] In 1974, the fraction represented
by the three categories most closely associated with computers and tel-
ecommunication—computers and electronic products, media and communications, and computer systems design—was only 35 percent. With
the computer peripherals boom in the early part of the next decade, the
share of these three sectors climbed to 62 percent in 1982; they then
peaked again at 61 percent during the dot-com boom in 2000. While
the shares of these three categories subsided somewhat in subsequent
years (in part reflecting the fact that social media firms are assigned
into multiple industry classes), they still remain far above the levels
seen in 1974.

In the United States, there are whole industries where venture capitalists have played a minor role—notably, real estate and health care.
Meanwhile, other sectors, such as media and communications, computers, and chemical products (which includes pharmaceuticals) garner a disproportionate share of venture dollars. The mapping between
venture financing and R&D spending is much closer.

These comparisons actually understate the differences between
venture portfolios and the economy as a whole in two important ways.
First, here we use broad boundaries to map out industries. Within
most of the categories, the venture funding is highly concentrated. For
instance, within the energy sector, venture funds have overwhelm-
ingly gone to renewable and "smart grid" technologies, rather than

conventional power-generation technologies. Second, firms within many categories where venture investors are active, such as social media, do not fit well within the conventional industry classification scheme used by governments in the United States and Europe. These companies are consequently split across several categories, making the venture funds look more diversified than they actually are.

What explains this dramatic increase in the concentration of industries represented in venture portfolios? The answer is simple: venture funds have done much better in certain categories than others. The groups specializing in some areas have had superior returns, which allowed them to garner more funds. Others have struggled or disappeared. As one illustration, groups whose initial mandate was in unpopular areas—such as Ampersand Ventures (with its original focus on advanced materials) and CMEA Capital (specialty chemicals and advanced materials)—were forced to reinvent themselves or perish.

No shortage exists of illustrations of the misadventures that venture groups have encountered when wandering into less-familiar sectors. For instance, venture groups have made many investments in the realm of apparel retailing and fashion. This fascination might appear peculiar on the face of it: most partners' wardrobes seem to prominently feature polo shirts embossed with the logos of firms in their portfolios and pleated khakis. And most of these investments appear to have ended catastrophically for all involved.

One of the most expensive illustrations of this point in recent years is Terralliance Technologies, an oil and gas exploration firm whose financing was led by Kleiner Perkins.[4] Terralliance was founded by Erland Olson, who had left NASA's Jet Propulsion Laboratory after fourteen years to cofound a firm making energy-efficient semiconductors for mobile phones and Bluetooth devices in 1998. After the firm was quickly sold to Broadcom for over $600 million, Olson began searching for another start-up.

Initially, Terralliance planned to use satellite data to search for water, but the team soon realized that the returns from finding oil would be far greater. Olson argued that the firm's technology—which sought to combine a wide variety of indicators, such as electrical transmissions from the soil and trace chemicals detected in the atmosphere—was far more reliable than the industry's mainstays, seismography and geology. By adopting his new technologies, he argued, the firm could leapfrog the stodgy oil companies.

This pitch was apparently compelling to Kleiner Perkins. The venture firm had enjoyed legendary success in information technology and biotechnology, having played a key role in the financing of Genentech, Sun, Netscape, Amazon, and Google. It had been undertaking a cleantech investment initiative for much of the 2000s. But at the time it invested in Terralliance in 2004, oil and gas was an unexplored sector for them.[5]

Kleiner played a key role in the first three, and progressively larger, financing rounds: $11 million in 2004, $35 million in 2005, and a whopping $250 million in 2006 (at which point the firm was valued at over $900 million). Over time, it sought to involve ever more investors, first Goldman Sachs, and then a wide variety of corporations and venture funds from as far away as Dubai. As one investor described it, "The real story, to boil it down, is as old as mankind: a charismatic individual with a compelling story you just want to believe . . . By force of personality, [Olson] convinced a lot of people. There's a real psychological story there. This was beyond numbers."[6]

Meanwhile, Terralliance had gone on a spending spree worthy of a Bruneian prince, with all the efficiency of the Keystone Kops. For instance, its costly well in Mozambique proved to be a dry hole; its well in eastern Turkey, a little more successful. The pièce de résistance, however, was undoubtedly paying the Ukrainian government $26 million for two Soviet-era fighter jets that were to be used to search for oil (but ended up not flying a single mission for Terralliance). When the questions about the firm's technology and governance proved too much for the Singaporean sovereign fund Temasek, which pulled back from a proposed billion-dollar investment, the firm rapidly went into a downward spiral (though not before Kleiner, Goldman, and other investors put another $54 million into the firm, most of which went to pay off other aggrieved investors who sued, alleging "numerous instances of egregious misconduct").

Ultimately, the firm would be renamed NEOS GeoSolutions and would raise another $60 million—led by (you guessed it) Goldman and Kleiner, now alongside Bill Gates and a Saudi oil-services company—in 2011. Maybe ultimately, the legendary venture firm will salvage one of its largest investments ever. But whatever the end game, Terralliance illustrates the dangers that even the best venture capitalists face when wandering away from the technologies they are familiar with.

This anecdotal account is corroborated by more systematic analyses as well. The most careful look at the returns from different classes of investment has been done by the consulting firm Sand Hill Econometrics, which uses data about individual transactions and the overall venture market to approximate the evolution of value of each firm.[7] It can "slice and dice" the overall performance of venture funds to get an approximate return for investments of each type.

We can see this if we look at the relative performance of all venture investments in each sector. (Of course, it would be very hard to construct such a portfolio, and the implicit assumption that one can buy and sell the holdings each day is false.) The performance of the information technology investments—whether hardware or software—is far greater than that of other sectors. A dollar invested in 1991 in venture-backed software firms would have turned (before the venture funds took their fees and cut of the proceeds) into more than $23 by 2011, for an annual return of close to 19 percent. A similar investment into the

bedraggled "other" category would have returned only $3, for an an-
nual return of just 6 percent. Once the venture capitalists' annual fees
(which typically run about 2 percent of the capital under management)
and profit share (20 percent or more of the capital gains) are factored
in, the performance difference would be even more egregious.

What lies behind this seeming limitation of the venture capitalists'
"pixie dust" remains an open question. Certainly, it is striking that
more and more of the funds have been concentrated in technologies
where the innovation cycle is shorter, such as social media and software, rather than long-gestating advanced materials and cleantech. Do
the huge financing needs of many firms in the energy and industrial
space defeat investors? Is this pattern just a consequence of historical
accident, where the groups gained valuable experience in a limited set
of areas? Or, as we will explore in chapter 8, are the commonly used
structures employed by venture funds to blame? Whatever the explanation, this represents an important limitation to the success of venture investors in the developed economies.

The Boom-Bust Venture Cycle

The venture market is extraordinarily uneven, moving from cycles of
feast to famine and back again. In some periods, far too many firms can
get access to financing, while in others, worthy companies languish
unfunded.

Funds operating in periods with little competition eventually experience very good returns, a pattern that may reflect the fact that the
funds operating during these years can invest in the most promising
firms at relatively modest valuations.[8] Over time, however, these high
returns attract the interest of institutional investors. What starts as
a trickle of funds ends as a torrent. The competition for deals rises,
as does the pricing of these transactions. Ultimately, the expansion

proves to be unsustainable, and returns fall. Then the cycle repeats itself all over again.

These cycles have led to considerable drama in the venture industry. Each industry downturn produces melodramatic claims that the venture industry is fundamentally broken, with too many investors competing for a limited supply of deals. For instance, in the dark days after the NASDAQ crash of 2000–2002, Steve Dow of the venerable firm Sevin Rosen indicated that his group was unlikely to raise a new fund. "The traditional venture model seems to us to be broken," he noted. "Too much money had flooded the venture business and too many companies were being given financing in every conceivable sector."[9] (More typically, the conclusion of the complaining venture capitalist is that everyone should exit the market except for the market observer and his best friends.)

This song has been repeated almost verbatim in every market downturn. "Dramatic inflows of cash weaken the 'fragile ecosystem' of the venture capital industry by forcing some to 'shovel' money into deals . . . The answer is to discourage more money from coming in and to suppress what [gets invested]," preached the *Venture Capital Journal* in 1993.[10] The same periodical bemoaned in 1980, "The rate of disbursements from venture investors to developing businesses continues to be extraordinary . . . [A] major limiting factor in expansion will be the availability of qualified venture investment managers. Direct experience is so critical to venture investment disciplines."[11] (In hindsight, the *Journal* was exactly wrong in both cases. The typical funds raised in the years of these two articles had a return of 26.1 percent and 21.6 percent, respectively, which remain among the two best vintage years for venture funds ever.[12])

Despite all the hype and drama, these boom-and-bust patterns are important ones. And the interest that these cycles have attracted is justified. It is natural to wonder why pension and other funds seem to almost inevitably put most of their money to work at exactly the wrong time. Why don't venture groups pull back from investing in market peaks, rather than continuing to dance the dance? While much remains uncertain about these cycles of boom and bust, several drivers of these patterns have been documented.

At least some of the deterioration of performance stems from the phenomenon of "money chasing deals."[13] As more money flows into their funds from institutional and individual investors, venture capitalists' willingness to pay more for deals increases: a doubling of inflows into venture funds led to an increase in valuation levels between 7 and 21 percent for otherwise identical deals. These results do not reflect improvements in the venture investment environment: when we look at the ultimate success of venture-backed firms, the success rates do not differ significantly between investments made during periods of relatively low inflows and valuations, and those of the boom years. But the findings, while suggesting how these cycles work, do not explain *why* they come about.

Part of the cyclic pattern in the venture activity stems from new funds.[14] During hot venture markets, many inexperienced groups raise capital. In many cases, these funds are raised from inexperienced investors, who are attracted by the excitement surrounding venture funds or by funds-of-funds, which target these investors. Often, they cannot get into top-tier funds, and instead reach out to less-experienced funds, not appreciating the differences across groups. (Figure 5-1 shows the disparity of venture fund performance. It depicts each fund as a separate observation, arranging them from the best to worst performers measured by net-of-fees internal rate of return. Only funds raised before 2004 are included, so that internal rates of return accurately reflect the funds' ultimate performance.) The performance of first-time funds tends to be weaker than that of other venture groups, and those of new funds raised in hot markets, particularly poor.

FIGURE 5-1

Internal rate of return, U.S. venture funds

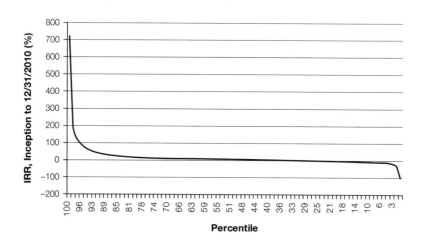

Source: Data from Thomson Reuters, "Venture Xpert Databases," http://wwwventurexpert.com.

Part of the deterioration in performance around booms reflects the changes in the venture funds.[15] Established groups often take advantage of these hot markets to aggressively increase their capital under management. (This decision is likely to be driven by the typical compensation that venture funds enjoy, which is largely driven by fees from capital under management.[16]) As venture groups grow in size, they tend to increase the capital that each partner is responsible for and to broaden the range of industries in which they invest. These changes are often associated with deteriorating performance.

Whatever the precise mechanisms behind these cycles, their impact on innovation is most worrisome. Skeptical observers of the venture scene frequently argue that these cycles can lead to the neglect of promising companies. For instance, during the deep venture trough of the 1970s—in 1975, no venture capital funds at all were raised in the United States—many companies seeking to develop pioneering personal computing hardware and software languished unfunded. Ultimately, these technologies emerged with revolutionary impact in the 1980s, but their emergence might have been accelerated had the venture market not been in such a deep funk during the 1970s.

Numerous other examples emerged during the recent financial crisis. Pegasus Biologics provides a particularly dramatic example.[17] Founded in 2004 and based in Irvine, California, it specialized in developing solutions for soft tissue repair. Originally focusing on using implants of collagen from horses to help heal ligament tears in the knee, it expanded its technology over time to address healing wounds of the arm and leg, devising treatments to forestall amputation.

Pegasus followed the typical "ladder" approach to financing: an initial angel round of $1.6 million in 2005, followed by $10 million from venture funds Three Arch Partners and Frazier Healthcare in 2005, and an additional $20 million from Onset Ventures, Affinity Management, and the two existing venture capital firms. As the financing rolled in, it expanded from four professionals to fifty-six full-time employees, ten contractors, and several hundred independent sales representatives. Revenues climbed from $3 million in 2006 to $9 million in 2008.

Pegasus's success was rooted in two factors. The first was that the company was founded at the right time. Biologics—the process of using a product created by a biologic process (such as a tissue, protein, or gene) as a therapeutic to treat disease, rather than one that was chemically synthesized—was attracting increasing interest from doctors and venture capitalists alike. Second, Pegasus's grafting, tissue regeneration, and wound-healing products experienced considerable success, rapidly gaining approval by the U.S. Food and Drug Administration and being extensively adopted by leading hospitals.

Everything seemed to be going in the right direction for Pegasus Biologics, but then came the Lehman collapse of 2008 and the ensuing financial crisis. Initially, the firm appeared to be untouched: demand remained strong, and in October 2008 the firm—under the leadership of a new CEO—announced ambitious plans to expand its sales force and manufacturing facility. But in early 2009, when one of the venture investors either would not or could not contribute additional cash for the next financing, things turned ugly. The remaining investors proposed a "cram down" round, wherein all investors would need to make a financial contribution or have their ownership stake severely diluted. After acrimonious debates, the plans for the new financing round collapsed. The firm was thrown into insolvency, its doors shuttered, and a sale of the firm's assets ensued.

In July 2009, Minnesota-based Synovis Life Technologies acquired Pegasus for the fire sale price of $12.1 million in cash. This was absolutely a golden opportunity for Synovis, which inherited the rich and diverse research portfolio from Pegasus, including patents, approved therapies, and inventory. Moreover, it was an excellent fit: Synovis used cow tissue to make products used, for instance, in hernia and breast reconstruction surgeries.

Pegasus was renamed Synovis Orthopedic and Woundcare. The subsidiary posted 181 percent growth in the first quarter of 2010 and began rapidly hiring new sales staff to fuel its geographic expansion. Because it had the resources—and the willingness to invest them—at a time when the venture investors were deficient in these qualities, Synovis was able to make what appears to be a tremendous investment.

During downturns like the recent financial crisis or the dot-com hangover, the differences between which ventures got funded by their existing investors and which ended up in distress like Pegasus were often capricious. Rick Townsend, in an intriguing analysis of the technology market collapse of 2000–2003, looks at the probability that firms in sectors unrelated to information technology (IT) during the collapse period got another financing round, and how this varied with their lead venture firm's exposure to the Internet sector.[18]

He compared non-IT firms whose backers invested heavily in Internet companies during the years leading up to the peak of the bubble with those whose financiers invested little in the Internet sector during that time. (Based on all observable characteristics, these firms were otherwise identical.) The unlucky ones with Internet-exposed backers were far less likely to raise another financing round. The analysis suggests that these unlucky firms—even though their technologies had nothing to do with the Internet, telecommunications, or software—experienced a 26 percent larger drop in the probability that they would raise additional funding than did those backed by funds without a heavy exposure to the Internet.

It might be thought that this termination of new ventures is not a big deal. After all, in the case of Pegasus, its technology was ultimately commercialized by Synovis. Similarly, the personal computing technology that may have languished unfunded during the 1970s saw the light of day in the next decade. But in addition to the delays inherent in this disruptive process—think of those people hobbling around on bad knees in 2009 who could have benefited from the Pegasus technology—there is also the question of its impact on incentives. If a potential entrepreneur realizes that even if he does everything right, his business may fail because he was unlucky in choosing a financier, his enthusiasm for the new venture may fade. He might well conclude that if he is going to be gambling, a trip to Vegas is a less costly and painful alternative.

Nor is the overfunding of firms during booms necessarily a good thing. These can lead to wasteful duplication, as multiple companies pursue the same opportunity, with each follower often being ever more marginal. The initial market leader's staff is frequently poached by the me-too followers, disrupting the progress of the firm with the best chance of success. Moreover, once the overfunding subsides, the firms that still survive struggle to attract funding, as the sector often takes on a poisonous atmosphere that deters venture investors.

Numerous examples of such crazed duplication can be offered: the recent plethora of social networking companies, or the frenzy surrounding B2B and B2C Internet companies in the late 1990s come to mind. One of the classic examples was during the early 1980s, when nineteen disk-drive companies received venture capital financing.[19] Two-thirds of these investments came in 1982 and 1983, as the valuation of publicly traded computer-hardware firms soared. During these years, equity financing from other sources—for instance, from corporate partners and the public markets, which welcomed initial public offerings from disk-drive companies—was also plentiful. Though industry growth exploded during these years (sales soared from $27 million in 1978 to $1.3 *billion* in 1983), industry analysts questioned whether the scale of investment really made sense, given the fierce price competition that almost inevitably would result.

The music came to an abrupt stop in mid-1983. Between October 1983 and December 1984, the average publicly traded disk-drive firm lost a whopping 68 percent of its value. After the abrupt decline in market valuations in late 1983 and 1984, obtaining financing became far more difficult for venture-backed disk drive firms as well. A brutal price war ensued, with thinly capitalized undiversified firms suffering the most: rivals selling products similar to those of these weaklings appeared to cut prices particularly dramatically. As a result, numerous disk-drive manufacturers that had yet to go public went out of business. Venture capitalists, suffering from poor returns, recoiled from the industry, making this period incredibly disruptive to all firms within the industry.

Mercurial Public Markets

Venture capitalists depend critically on the public markets, due to the nature of their investors. They invest in venture funds—whether a Middle Eastern sovereign fund, an Ivy League endowment, or a state pension fund—for one primary reason: to make good returns. And without being able to exit their investments, venture capitalists will be unable to return the liquid securities (whether cash or publicly traded stock) that these investors need to heat buildings or pay pensioners.

There are two primary ways that venture capitalists exit successful investments: by taking the companies public or selling them to corporate buyers.[20] The volumes of the two types of exits have varied over the years, but it is clear that they move together. When public markets are hot, it is typically relatively easy to arrange IPOs. Even if the venture fund decides to sell a company in its portfolio to a corporation, the presence of a high-priced alternative exit route will ensure an attractive valuation.

For instance, when the investment team led by diversified technology investors Silver Lake Partners and venture firm Andreessen-Horowitz were approached by Microsoft in 2011 regarding the possible purchase of Skype (which they had bought from eBay for $2.75 billion in 2009), the fact that the private firm had already extensively explored going public shaped the conversation. During their conversations with investors, Skype's managers and backers had learned that the bankers anticipated a valuation of $7 billion given the market's enthusiasm at the time for web-related companies.[21] Microsoft ultimately ended up paying $8.5 billion for the acquisition. The state of the public markets seems to drive the pricing of and the ability to undertake exits of all types.

The reliance on the public markets would not be a problem if these markets could be trusted to appropriately reward the pursuit of innovation and other long-run objectives. But a myriad of accounts and studies have questioned whether this is the case in the public markets. In Jeremy Stein's classic theoretical exposition, managers want to do the right thing, by making long-run investments.[22] He points out that if the institutional investors in the market knew everything managers did, there would be no potential problem. But in his model (as in the real world), public market investors are not as fully informed about the firm's prospects as the managers are. As a result, they are likely to become nervous if profits do not rise in keeping with expectations. Managers realize that if this scenario ensues, corporate raiders may swoop in and buy up the undervalued stock. Eager to avoid such a scenario, the managers may spend less on R&D than they should. Such a strategy should keep earnings high, investors happy, and raiders at bay. But in the long run, innovation at the firm will suffer. While this is just one formulation, one could imagine numerous scenarios where managers would rationally make decisions that are not optimal.

How common this scenario is in the real world is a topic of lively debate. While there are certainly examples of operating firms that made inappropriate decisions due to public market pressures, an even more dramatic and relevant example is the investment group 3i.[23] The oldest private equity group in Europe, 3i stood out from its peers along several dimensions. It had been founded in 1945 by the U.K. government and funded by a consortium of banks to provide capital for small and medium-sized businesses in the post–World War II rebuilding effort. 3i initially used both debt and equity to fulfill its mandate, and expanded into far-flung product lines, like consulting, securities underwriting, and ship financing. Over time, it became a more traditional private equity group, focusing on buyouts and growth capital and shedding many of the ancillary activities. It also expanded its footprint into continental Europe and elsewhere. These efforts yielded some real successes, including investments in Bond Helicopters, Caledonian Airways (later British Caledonian), and Oxford Instruments, the pioneer of magnetic resonance imaging.

In July 1994, 3i made a transition that was virtually unheard of at the time: the firm went public on the London Stock Exchange. In this way, the banks could unwind their shares and 3i could attract a new investor base. The offering was initially seen as a success: by

September of that year, 3i's market cap had risen enough that it was included in the FTSE-100 market index of leading British firms.

As the decade progressed, however, 3i's management became increasingly frustrated with its perception and valuation in the market. The market viewed the firm as, in the words of one observer, more of a "quaint throwback than a go-ahead investor."[24] As the technology market had soared, 3i appeared to have been left behind: its stock price had increased only 39 percent between the beginning of 1996 and 1999, while the tech-heavy NASDAQ had more than doubled and even the staid FT-100 firms had grown by nearly 50 percent.[25]

At the behest of Brian Larcombe, who became 3i's CEO in 1997, the firm sought to address the market perception. Beginning in 1999, the investment company began aggressively moving into early-stage investments. 3i increased its pace of start-up investments tenfold in three years, and shifted its later-stage investing from traditional transactions involving mature firms to follow-on rounds in technology companies. Meanwhile, in 1999 it opened offices in Silicon Valley and Boston to increase its exposure to early-stage deals there. The result was undoubtedly successful—in the short run. Its stock price nearly tripled between January 1, 1999, and mid-2000, breaking the thousand-pence mark in August 2000.

But dark clouds were gathering. In March 2000, the market for new offerings of technology companies—and the valuation of those firms that were traded—began collapsing. Many seasoned venture capitalists, having been through such market cycles before, began pulling back from new investments, instead focusing on nurturing the existing companies in their portfolio. Meanwhile, 3i, convinced it was giving the market what it wanted, plunged ahead. In mid-April 2000, it launched a £400 million European technology fund quoted on the London Stock Exchange. Its investments in start-up firms actually peaked in the fiscal year ending March 31, 2001, a year in which the technology stocks (and venture returns) had been in a virtually continuous state of decline. Even as late as May 2001, the firm expressed its desire to keep investing at least half its new capital into the technology sector. Larcombe professed: "This is not a time to be shy of making new investments— I have little doubt that the fastest growing businesses will continue to be technology companies."[26]

By the end of 2001, however, the optimistic tone was gone at 3i. The extent of the folly was apparent. The firm reported a loss of almost £1 billion for the fiscal year ending March 31, 2002, driven by the poor performance of its technology portfolio. Needless to say, its stock price plummeted as well, falling below 300 pence by the first quarter of 2003: it had given back all the gains it had made in the giddy years of 1999 and 2000, and was trading below where it was at the beginning of 1996. The firm responded by dramatically curtailing its new technology investments, laying off 17 percent of its staff, and entering a "triage" process to salvage its portfolio.

But the biggest cost to the firm may have been the distraction in the ensuing years posed by this troubled portfolio. The process of working out the portfolio of doomed technology companies proved to be a tortuous one: it was not until 2008–2009 that 3i closed down its Silicon Valley office and sold its remaining holdings to a consortium of secondary funds. While management struggled with these issues through the early and mid-2000s, opportunities abounded in 3i's traditional market: equity investments in midsized European firms. And while 3i's later-stage funds showed attractive returns during this period, younger rivals such as Montagu and Permira had superior returns and more rapid growth. It is hard not to feel that the distraction of dealing with the portfolio of troubled technology investments—brought about by a desire to please the public markets—led to 3i's inability to fully take advantage of the opportunities when conditions in its core markets were best.

A number of studies have offered illustrations of venture capitalists—and the firms they back—acting in ways that appear to be driven more by a desire to impress investors than to create value in the long run. For instance, papers by Paul Gompers, Peggy Lee, and Sunil Wahal look at what they term "grandstanding" by young venture capital firms: that is, just like the showboat wide receiver strutting down the sideline after catching the football, they take unproductive actions that signal their ability to potential investors.[27] Specifically, they argue that young venture capital firms bring companies public earlier than they should, in an effort to establish a reputation and successfully raise capital for new funds.

To demonstrate this, they look at venture-backed firms that went public, comparing those taken to market by seasoned and inexperienced venture funds. Companies backed by less-established venture capital firms are younger (by nearly two years) at the time they go public, and in most cases, the venture investors have had a shorter relationship with and own a smaller stake in these firms. These offerings are more likely to experience a substantial rise in price on the first trading day as well, which they suggest means that the only way the underwriting banks could cajole investors to buy these shares was by offering them a substantial discount. At the same time, younger venture capitalists benefit considerably from taking companies public: while investors are typically confident in the track record of more established groups, the validation of an IPO reassures them that a new group is on the right track. Young venture capital firms, eager to raise another fund, continue to rush young firms to market, despite the fact that such premature offerings are harmful to the companies and to the returns the venture capitalists' investors will garner.

Public markets have two faces. They provide an essential way that venture capitalists can exit their investments or determine a fair price for the firms they are selling to corporate buyers. Without this liquidity, few institutional or individual investors would find venture capital to be an attractive area.

At the same time, public markets appear to be prone to misconceptions. Their perception of the appropriate strategic course may become distorted, as they are swayed by fashion or simply cannot understand the specifics of a firm's situation. A young company and its financial backers may be unwilling or unable to resist these imprecations. As a result, everyone may suffer: the investors may lose money, the young firm may be unsuccessful in developing its promised new products, and the economy as a whole may be less innovative and dynamic.

Uneasy Oversight

Venture investors typically have many tools through which they can steer the companies that they fund, from preferred stock with special privileges to board seats to the right to replace the founder as CEO. But any successful venture-backed start-up must be a partnership between the financiers and the entrepreneurs who ultimately do the hard work of running the firm. Even if the venture capitalist desires to micromanage every company in the firm's portfolio, the limits of time and attention will lead inexorably to a reliance on entrepreneurial managers.

Entrepreneurs, to be sure, have made wonderful contributions to the world. From Henry Ford to Mark Zuckerberg, our world has been transformed—and our nation enriched—because of the vision and focus of these leaders. But alongside these great accomplishments often comes a kind of egotism and paranoia that can lead to value-destroying behavior, a pattern as well documented in the novels of Charles Dickens and Honoré de Balzac as in issues of *BusinessWeek* and *Fortune.*

Economists rather drily refer to these tensions as involving "private benefits of control." The entrepreneur may choose actions that gratify him- or herself rather than create value for the firm as a whole. The most obvious examples are outright wasteful expenditures, such as the launch parties thrown by dot-com entrepreneurs during the late 1990s that became sufficiently notorious to merit their own copiously documented entry in Wikipedia.[28]

Even more pernicious are decisions that seem reasonable at face value, but ultimately take the firm in the wrong direction. Frequently, these take the form of entrepreneurs seeking to go head-to-head with established players, and become industry giants themselves. But often to do this, the firm must make huge, risky investments in developing manufacturing skills, a professional sales force, and the like: complementary assets that are often very challenging to build. In many instances, as Joshua Gans and Scott Stern discuss, the fledgling firms would have been far better off earning returns in the "market for ideas": licensing their concepts to larger firms or simply selling the start-up in its entirety.[29] Whatever the motivation for the decision to go it alone, such episodes are sufficiently commonplace that my colleague Noam Wasserman has tagged them "choosing to be king."[30]

Nowhere is this tendency more evident than in the biotechnology industry, whose history is littered with the carcasses of start-ups that sought to take on Eli Lilly or Novartis mano a mano. Ilan Guedj and David Scharfstein provide systematic evidence of the costs of such behavior.[31] They look at young biotechnology firms specializing in cancer treatment that have not yet garnered any revenues from product sales, and contrast them with established, typically much larger pharmaceutical firms. After a drug compound has been identified through preclinical research, the biggest investments that biopharmaceutical firms must make are the clinical trials they must conduct to prove the safety and efficacy of a potential new drug: the U.S. Food and Drug Administration requires three rounds of tests of increasing stringency. They hypothesize that the small firms, which typically go public long before they have revenues (much less profits), will be tempted to "cut corners" in these trials in order to have a good story to tell investors and keep their valuations high. More mature firms, with a proven track record of success and a pipeline of alternative drug candidates, should be less likely to continue to pour resources into a borderline project.

And this is exactly what they find. Early-stage firms, particularly those that have successfully raised large sums from the market, are much more likely to push drug candidates from the initial phase I to phase II trials (61 percent of the start-ups' drugs move to Phase II within a two-year period vs. 45 percent of the pharmaceutical companies'). Yet once the young firm moves its drug candidate into the more exacting later-stage trials, things turn ugly. In the phase II trials, the likelihood that the tumors shrink—a key indicator of success—is less than half the rate for the proposed drugs of the newer firms than for those of the larger corporations. Similarly, the probability of moving into the final phase of trials is much smaller for the biotech companies (14 percent as opposed to 35 percent for the larger firms). These findings suggest that the desire to impress the public markets leads young firms to plunge ahead with marginal drug candidates, when they and their investors would have been better served had they abandoned the projects early.

Were entrepreneurs totally rational and unbiased decision makers, many of them would probably have never made the plunge to found their companies in the first place. (See the calculations of Robert Hall and Susan Woodward, for instance, who conclude that if an entrepreneur has even a modest dislike of risk, the returns of starting a new venture have historically been extremely modest.[32]) These characteristics, while providing fuel for the entrepreneurial engine, can also lead start-ups to make value-destroying decisions. Although venture capitalists can try to rein in the tendency of entrepreneurs to seek private benefits—usually by "doubling down" on a troubled technology—all the control mechanisms in the world will have only limited effectiveness. Ultimately, these dynamics are likely to lead to many costly "dry holes" in the quest for venture-backed innovation.

Final Thoughts

Senior partners at an established venture firm are likely to have a pretty sanguine view of their own (and their partners') ability to effect positive change in the firms that they fund. This is understandable: one is unlikely to be successful at committing skittish institutions' money to nascent start-ups without a considerable degree of self-confidence.

But the venture capital model is no panacea. Even the most established firms have had to adapt to various realities: the modest success of venture investments outside a relatively narrow band of technology; the impact of booms and busts, where funds from investors are either in verdant oversupply or very scarce; the mercurial public markets; and the limitations of shaping entrepreneurial actions. All the factors limit the degree to which venture capital can be a spur to innovation.

It is natural to wonder, though, whether this model and the corporate research laboratory explored in part 1 can benefit from each other's approaches. As we have seen, each institution's strengths and weaknesses seem quite complementary. It is this challenging question we will explore in part 3.

Part Three

The Best of
Both Worlds?

6

R&D, Meet VC: The Promise of Corporate Venturing

A CONVERSATION WITH a typical venture capitalist will reveal mixed opinions, at best, about corporate venturing—and in some cases, a distaste that is rivaled only by a mob boss's view of a former colleague in the witness protection program. Bob Ackerman of Allegis Capital has depicted the corporate venturing process in grim terms:[1]

> The challenge is that corporations just show up and four guys get out of the car with their corporate tee-shirts and singing the company song. They have one perception of the size and power of their business, but start-ups and their investors see it differently. Often we view them as the dinosaurs we're trying to kill, the market opportunity we're trying to capture . . . So they get used as cannon fodder. It's just the nature of how things get done. I don't pretend that it's easy. There's so much history of

venture capitalists abusing corporations and the corporations saying "it's different this time" when it's not.

But by adopting some of the approaches employed by venture capitalists, corporations *should* succeed in responding to dramatic technological change quickly, collecting insights that would take them many years to gain otherwise. The start-ups should benefit as well from the larger firms' intellectual resources and deep pockets.

To understand the unease—not to mention the contempt—with which many venture investors view corporations seeking to interact with start-ups, it is helpful to look backwards, at the history of corporate venturing and in particular at two flaws: failures to think clearly about goals and to design appropriate compensation schemes.

But when we take stock of the more objective assessments of these efforts to mix and match these two modes of innovation, the view is surprisingly hopeful. The dubious reputation of these programs seems to be largely if not wholly unjustified. While there certainly have been many failures, corporate venturing programs have also succeeded in generating attractive returns and in advancing corporate strategic goals. Looking over the evidence, it is hard to conclude that there is no successful middle ground between the two models.

The Case for Corporate Venturing

A first—though perhaps less than persuasive—argument for why corporate venturing makes sense is its ubiquity. Again and again, corporations have turned to venture investing. There have been at least four waves of corporate activity in the venture capital arena: in the late 1960s, the mid-1980s, the late 1990s, and the early 2010s.

When we look at the early days of corporate venturing, when data on these investments was scant, it is only feasible to follow the announcement of the inception of such programs by major firms.[2]

These efforts were launched frequently during the late 1960s and early 1980s, which not surprisingly corresponded with two of the earliest booms in venture capital investments and venture-backed IPOs. In more recent years, there is much greater data on the extent of financing rounds with corporate ventures and on the dollar volume of the corporate investments. Recent years have not rivaled the dot-com bubble period in terms of the number or dollar volume of corporate investments: the year 2000, which saw over $15 billion in corporate venture investments in almost two thousand separate transactions, remains the undisputed high-water mark. But the corporate share of venture investments has climbed steadily. In fact, in the first half of 2011, corporate funds invested almost 10 percent of every venture dollar, a level exceeded only between 1999 and 2001 (when corporate activity peaked at 15.4 percent).

This enduring fascination might suggest that these programs contribute something valuable. A cynic would note, however, that many other dubious fads—from matrix management to low-carb diets—have appeared again and again, despite the lack of much evidence of their effectiveness.

A more convincing argument for such hybrid initiatives can be made based on case studies. For instance, the pharmaceutical industry has seen numerous corporate venturing programs that have allowed the firms to catch up to the latest scientific advances. These firms have had to deal with a rapid shift in technology, which in many cases has rendered their traditional skills and substantial assets in chemistry-based research far less valuable than they once were. One recent example of a firm that has sought to address these changes with a corporate venturing program is Eli Lilly.[3] The firm began a corporate venture initiative in 2001, setting up a unit to invest in strategically relevant areas with an eye to generating attractive financial returns. The firm correctly anticipated that the venture arm's ability to bring the resources of the corporate parent to bear would make it an attractive partner and would give Lilly an early look at promising technologies. The venture unit's structure, particularly the decision to have the venture capitalists be corporate employees without any profit share, led to a steady stream of defections. Ultimately, in 2009 Lilly Ventures was reconfigured as a freestanding fund that retained its financial and strategic mandates.

Intel turned to corporate venturing for an entirely different reason: to stimulate demand for its semiconductors.[4] In late 1998, the company's venture unit, Intel Capital, sought to establish a fund that would help diffuse its next-generation semiconductor chip into the market. By investing in many firms (often competitors) that were developing hardware and software that capitalized on the new chip's power, the fund managers believed they could accelerate the launch of the chip and thereby enhance Intel's profits. (The bulk of the profits from new chips comes from sales during the earliest months and years after the product launch, before rivals have a chance to introduce competing offerings.)

This catalytic role has since been expanded to other technologies. Intel Capital played a key role in seeding companies developing wireless Internet products around the 802.11 technology, which Intel had aggressively championed. Reflecting these priorities, Intel Capital traditionally took a passive approach, relying on the independent venture funds that it syndicated almost all its deals with to provide the bulk of the oversight and strategic guidance to the firms. It also made investments in technologies that complemented its product portfolio, but which it did not intend to manufacture itself. In May 2011, the firm invested its ten billionth dollar in a new venture, cementing its status as one of the most established venture groups of any type.[5]

Analog Devices' historical foray into venture capital represents yet a third motivation: to understand a competitive threat.[6] The company specialized in developing silicon-based, or CMOS, semiconductors, which dominated the industry at the time. During the early 1980s, some players in the industry searched for alternative technologies—such as gallium arsenide (GaAs) and bipolar semiconductors—to go head to head with CMOS technology. This company, in response, ran a corporate venture program, Analog Devices Enterprises, from 1979 through 1985 that invested in these competing technologies. The returns from its portfolio were very poor: only one of its thirteen portfolio companies went public, and it did so after so many financing rounds that Analog's stake was highly diluted. The corporation wrote off more than half the $26 million it (in conjunction with its partner, Amoco) originally invested in these firms. But this seeming failure was paradoxically a success. Over time, these technological threats proved much less formidable than people in the industry initially believed: it was far harder to make competitive GaAs semiconductors. Accordingly, the valuations assigned to CMOS-based manufacturers spiked: Analog's value, for example, increased sevenfold during these years. Analog used its corporate venturing program as an insurance policy. Granted, the policy did not pay out—but just as with a homeowner who gladly renews his fire insurance year after year, this is good news.

So corporate venturing does fulfill three needs, at least. The first of these is the ability to *respond quickly* to changing circumstances. While internal research laboratories can be time consuming to build up—particularly the identification and recruitment of the right people—corporate venture programs can often quickly identify suitable firms in a promising area. For example, five months before the 2003 introduction of its Centrino chip set (which had wireless capability installed alongside the computer's central processing unit), Intel Capital announced its intention to invest $150 million into companies that were promoting the adoption of Wi-Fi networks.[7] Even before the new product had launched, the corporate venturing arm had invested in Cometa Network, a developer of public Wi-Fi hot spots, and Pronto Networks, which operates management systems handling functions such as security and billing for large network providers. Had Intel tried to develop these capabilities in-house, it would doubtless have taken them far longer to break into these unfamiliar businesses.

A second key benefit is the ability to leverage outside funds. Essentially, this gives the firm a much greater bang for the buck than if they were spending their own funds entirely. This aspect is particularly important when there is considerable technological uncertainty, as the Analog Devices case illustrates. Had Analog instead made a major shift in its research laboratories to focus on GaAs and pursued each of the options that its portfolio companies were, the cost—in terms of both dollar expenditures and disruption of ongoing research efforts—would have undoubtedly been far greater than their losses from the corporate venturing investments. Because of this leverage, the corporate venturing program proved to be a cheap way of garnering critical strategic information.

A recent dramatic example of the use of corporate venture capital to leverage corporate investment was the iFund.[8] This fund was launched in March 2008 by the venerable venture group Kleiner Perkins, with Apple's involvement and support, on the same day that outside developers were allowed to begin working on sanctioned iPhone applications. The $100 million fund—subsequently doubled in size—sought to invest in companies that were developing games, tools, and other iPhone (and later iPad) features. In this way, Apple was able to rapidly build a critical mass of applications for its new phone while spending very little itself. (The contrast with Apple's rival Nokia, who eschewed such an approach when promoting its now-abandoned rival Symbian system, is dramatic.) Given the success of the iFund, it is not surprising that similar efforts have been launched by, among others, Research In Motion (to encourage the development of third-party applications for the BlackBerry) and Facebook (who teamed with Kleiner, Amazon,

Zygna, and other tech luminaries to establish the sFund, devoted to promoting companies that work with social media sites).

A final advantage of corporate venturing is its ability to quickly change course. In many cases, firms—for all the talk of R&D portfolio management—find it difficult to abandon internal projects. Given these difficulties, the arms-length relationship between the corporation and the ventures it backed has real advantages. Even if the corporation is itself unwilling to pull the plug on an unpromising initiative, the presence of coinvestors may force such a decision.

Corporate venturing should also allow firms to avoid many of the pitfalls affecting freestanding venture funds. Corporate venture capital funds should be able to succeed in a broader range of technologies the independent funds have, due to the skills resident in the corporate parent. The deep pockets of the parent should mean that investments are less dependent on the ebb and flow of fund-raising cycles. And the fact that many of these firms may ultimately be purchased by the corporation should imply fewer distortions introduced by public market cycles.

This theory sounds compelling, which might explain why in recent years we have seen companies as diverse as Google, BMW, and General Mills launch corporate venturing efforts. But the implementation of these efforts has not always been smooth.

What Can Go Wrong

Perhaps not surprisingly, the same two critical issues that corporate R&D programs have struggled with have also affected corporate venturing initiatives. First is the confusion over the ultimate objectives of the research program, which translates into poor decision making and swings between an emphasis on central and divisional research. In many cases, confusion over goals has also characterized the design of corporate venturing efforts. The second problem is an unwillingness to match—or even come close to matching—the compensation offered by independent venture groups.

The classic case of Exxon Enterprises, whose venture capital effort ranks among the earliest and most spectacular failures in the field, provides a vivid illustration.[9] The oil giant (called Esso at the time), seeking to diversify its product line, had launched its venture program way back in 1964.

The program showed all the hallmarks of an effort without a consistent strategic direction. It began with a mandate to exploit technology in Exxon's corporate laboratories; for example, making building materials out of petroleum derivatives. In the late 1960s, however, the fund managers decided to make minority investments in a wide variety of industries, from advanced materials to air-pollution-control equipment to medical devices. This flurry of investments by the corporation was made just before the market went into an extended swoon. In the late 1970s, the strategy changed yet again: the program now focused solely on systems for office use, with a particular focus on advanced computing. Finally, in 1985, Exxon abandoned the venture effort entirely. Each shift in corporate strategy had brought on waves of costly write-downs. The information-systems effort alone generated an estimated $2 billion in losses for the corporation.

What explains this disaster? In part, the lack of a clear mandate from the top was at fault: senior management at Exxon could not agree on the program's overarching purpose. But there was considerably more blame to go around. The corporate venture team came to the project with scant investment experience and made numerous poor decisions. Moreover, various divisions at Exxon insisted on detailed reviews of the program, which were designed with an eye toward shifting the program's direction in ways that the various divisional leaders deemed most beneficial for their interests. These reviews consumed so much time that they distracted the fund managers' attention away from the selection and oversight of investments. Meanwhile, various organizations within the corporation had a hand in structuring the program. For instance, Exxon's human resources staff complained that the venture firms' compensation schemes did not mirror those of the overall corporation. In the late 1970s, the personnel group succeeded in replacing the venture staff's separate stock-option schemes with a standard salary-plus-bonus plan. An exodus of fund managers soon followed.

How do smart companies go so wrong when it comes to designing corporate venturing programs? The confused objectives that characterize many programs may reflect the realities of getting approval of a new initiative. If we think about a traditional independent venture fund, the goal is simple: to maximize the financial returns of the limited partners while behaving in a reasonably ethical (or at least legal) manner. All the limited partners of a typical fund, whether an endowment, pension, or sovereign fund, would heartily agree that high returns are desirable.

Now consider the situation of the champion of a corporate venturing program. To get the approval to set up the program, he or she will probably have to satisfy a wide range of constituencies. And to get these approvals, more features are often added to the program: for instance, a target to identify potential acquisitions to please the business development group, an objective about financial returns to satisfy the chief financial officer, and a goal of getting a window on emerging technologies to please the R&D department. By the time a program has approval from all relevant parties, it often resembles an overladen Christmas tree, with boughs bending under the weight of all the ornaments that have been added. Almost inevitably, the task that then faces the group responsible for managing the program is a virtually impossible one. The program is doomed from the beginning.

The very process of making decisions can lead corporate programs to run into issues with venture market cyclicality. Instead of escaping the boom-bust patterns that many investors follow, corporate groups often mirror these patterns. In many cases, the process of getting approval to begin a corporate program is so arduous that these ventures are launched only when the market is attracting a frenzy of interest from independent groups—and lots of media attention, which attracts the attention of senior management and board members. As we saw, these times tend to be the worst ones for investing: valuations are likely to be the highest, business plans the most problematic, and ultimate returns the lowest. The same dynamics may lead corporate venture programs to focus on the technologies that are most popular among private investors. Corporate venture investments in 2010 and 2011 tended to be concentrated in the same areas as those of traditional funds: biotechnology, software, and energy (though the latter is considerably overrepresented in corporate venturing, at least when it comes to dollars invested).[10] This pattern may reflect pressures to avoid taking contrarian approaches that might lead to the identification of neglected opportunities. (Alternatively, it may simply reflect the difficulties of doing venture investments of any kind outside of a few industries.)

One frequent manifestation of these confused objectives is a difficulty in killing projects. While the very finite time frame of independent venture funds and pressures for a high rate of return from investors often lead traditional funds to be ruthless in terminating struggling firms, far too often ventures with modest prospects backed by corporations (and projects in corporate laboratories more generally) are allowed to linger. Although we don't fully understand why this is so, the complex objectives that often characterize hybrid corporate initiatives are likely partially at fault. The distraction, cost, and complexity these slowly failing projects introduce is substantial.

The failure to offer adequate compensation to corporate venturing groups also mirrors the challenges that corporations face when rewarding innovators within their labs. For instance, when Lilly Ventures was hit by its initial wave of defections, it benchmarked its compensation levels against those of independent firms. The conclusion was that only the most junior staff were being rewarded at near a market level. Yet the corporation's senior management and human resource professionals resisted changing the scheme, pointing out the high quality of junior hires that the firm was making.[11] It was not until 2009 that the firm's management agreed to the restructuring.

In some cases, rich rewards for corporate venture investors are not needed. For most of its history, Intel's fund has emphasized making passive investments in a wide variety of companies in selected categories, analogous to a mutual fund following an index fund approach. Such a program may place much less of a premium on the skills of the investment team, and the rewards can be scaled down accordingly.

But in most programs, the demands on the corporate venturing team are considerable. These investors are asked to carefully assess portfolio firms, attend board meetings, and provide strategic guidance, just like independent venture capitalists. In some respects, their job may be easier: typically, the corporation provides all the funds, sparing them from having to undertake the arduous quest for capital. But in other respects, their jobs are considerably harder. Not only are corporate venture investors responsible for managing the tricky interface between entrepreneurial firms and an often slow-moving corporation, but they are also typically asked to be their firms' ambassadors to the venture community.

The many corporations that have eschewed incentive compensation have had to face a steady stream of defections once the junior investors have mastered the venture process. After too many board meetings in which the corporate investor parks his Fiesta next to the independent venture capitalists' Ferraris, the temptation to go else-where becomes too great. The corporation, having borne the cost of training the fledgling venture investor, does not get to benefit from the harvest.

These issues also manifest themselves when it comes to rewarding the managers of spun-out firms. Often, firms resist granting substantial equity stakes to these corporate entrepreneurs. Even more troubling, all too frequently the management teams are pushed to accept sketchily defined "shadow equity" rather than a real claim on the new entity. Much of the corporate resistance stems from the fact that their entrepreneurs are commercializing a technology that belongs to the company. This attitude is shortsighted, as it neglects the fact that there is typically a long road between a promising technology and a viable product. Without an adequate share of the upside, corporate entrepreneurs are often tempted to look elsewhere.

Many of the programs with the greatest stability—in terms of both management team and mission—have been characterized by high-powered incentives. An example is SmithKline Beecham's S.R. One, which operated under a single head, Peter Sears, from 1985 to 1999.[12] Not only was the management team stable, but the fund achieved impressive successes. For instance, it invested in biotech firms like Amgen, Cephalon, and Sepracor, and coinvested with major venture firms such as Kleiner Perkins and New Enterprise Associates. Its compensation scheme played a large role in this success. During most of this period, the corporate venture capitalists received 15 percent of the profits they generated, as well as a bonus based on less-tangible benefits to the corporation, which could run as high as another 5 percent of the fund's capital gains. This approach kept SmithKline's venture investors sensitive to both its financial objectives and the parent company's strategic needs.

The large-sample evidence also suggests that these incentives really do matter. Gary Dushnitsky and Zur Shapira survey corporate venture groups to understand the incentive schemes they employ and then relate these to their investment behavior.[13] The groups with higher-powered incentives are more likely to undertake investing akin to traditional venture funds—for instance, investing in earlier-stage companies. Moreover, the better-incentivized groups are more likely to have exited their transactions through an IPO or acquisition.

Incentive schemes may lead to changes in the corporate investors' behavior, as the authors argue. Or it may be that firms—anticipating that their program will require intensive involvement by investors—attract such venture capitalists by offering these kinds of incentives. (It would be hard for them to definitively answer this unless they had been able to find a broad-minded company who would be willing to let them run experiments with their venture groups' pay schemes over the next decade or so.) But in any case, a strong link between incentives and performance seems to be present.

Taking Stock

Economists have taken two approaches to assess the extent to which firms have been able to successfully introduce such a hybrid form of investing. The first has been to look at the success of the investments themselves—whether the portfolio firms went public or were acquired for an attractive valuation. This view is limited, because it focuses on financial returns, and many of the goals that corporations have when launching these programs are strategic ones. The alternative approach is to look carefully at the corporations' and start-ups' activities, to discern whether they have benefited from their interactions.

Perhaps looking at the financial returns from corporate investments is not totally misguided. Daniel Kang and Vik Nanda looked at seventy-one biopharmaceutical firms that had corporate venturing programs between 1985 and 2005, and sought to assess what the returns from these initiatives were, whether financial or progress toward development of new drugs.[14] These two sets of returns go hand in hand: if a promising investment goes belly-up, the corporation will get zero financial return and is unlikely to get much strategically useful information either (unless it learns what doesn't work, as in the Analog Devices case discussed above). The firms that made financially successful investments also experienced more success in drug development. Their analysis suggests that if the studies find positive returns of one type, we can be more confident that the other class of benefits is also present.

Paul Gompers and I examined the financial returns from over thirty thousand investments into entrepreneurial firms by venture capital organizations, comparing investments by two types of funds: independent venture partnerships and corporate funds.[15]

When we looked at the success of the investments, the results were surprising. Far from being outright failures, corporate venture investments in entrepreneurial firms appear to be at least as successful (using such measures as the probability of a portfolio firm going public) as those backed by independent venture organizations. For instance, 35 percent of the investments by corporate funds were in companies that had gone public by the end of the sample period, as opposed to 31 percent for independent funds.

But the pattern of success is not uniform. The success of the corporate programs is particularly pronounced for investments in which there is strategic overlap between the stated focus of the corporate parent and the business of the portfolio firm, as determined by a careful reading of the firm's annual report in the year before the investment. The probability of going public by the end of the sample period, for instance, is much greater when there are such similarities than when no overlap appears. Similarly, corporate programs without a strong strategic focus are much less stable than those of independent funds, typically ceasing operations after only a few investments, while those with a strong strategic focus are almost as long-lived as independent funds.

Another approach is to look at the progress of firms that get corporate venture funding: we would be much more confident that the corporate parents were benefiting from these initiatives if the firms in their portfolios proved to be special. Thomas Chemmanur, Elena Loutskina, and Xuan Tian seek to understand the experiences of firms backed by corporate venture investors, comparing them to those backed by traditional venture investors.[16] The patterns they find are striking. The firms backed by corporate venture capitalists are more productive in filing patents. For instance, in the first four years after going public, the corporate-backed firms have 47 percent more awards of equally high quality than those backed by independent firms. Moreover, the market appears to recognize the superior value that corporate investors add. When companies backed by corporate venture funds go public, the firms benefit from the greater participation by high-quality market players—whether investment banks, equity analysts, or institutional investors—more than the typical venture-backed offering, and stock prices outperform offerings of both those backed by traditional venture groups and the market as a whole in the five years after going public.

Consistent with the findings regarding financial returns discussed above, the entrepreneurs patent more when there is a good strategic fit between the corporate financier and the start-up. Similarly, "tolerance for failure"—which is measured as the amount of time that the various venture teams let companies that ultimately died linger before terminating them—matters. While letting unpromising firms linger for too long is counterproductive, at the same time, the uncertainties associated with the start-up process are such that not being too "trigger happy" is also important: many ultimately successful firms have dark moments along their journeys. Corporate venture investors have typically been more willing than independent groups to give even ultimately unsuccessful firms more time, and this quality is associated with more innovation. The authors suggest that the corporations' deep knowledge about the start-ups' industry and longer-run perspective explains the superior performance.

Yet another body of work has looked at the success of firms that have spun off from larger concerns, including both start-ups funded by corporate venturing programs and less amicable separations. A series of industry studies—especially the work of Steven Klepper on the automotive industry and more recently Ronnie Chatterji on medical devices—has highlighted the success that spin-outs from established players have enjoyed in new industries.[17] A cross-industry study by Paul Gompers, David Scharfstein, and myself suggests these spin-outs disproportionately arise from established firms that are themselves more entrepreneurial: for instance, those that are based in Silicon Valley and were originally venture backed.[18] While this evidence is less direct, it again suggests the innovative potential of would-be spin-outs from larger concerns, which a corporate venturing initiative may bring to life.

Taken together, the evidence suggests that corporate venturing programs can play a valuable role. In particular, despite the pitfalls that have contributed to program instability, such as the lack of a clear mission or inappropriate compensation schemes, the overall impact is quite positive. Whatever limitations individual programs may face, the expertise and the longer time horizon that the corporate investor brings to the table typically create real value.

A Related Avenue: Alliances

While corporate venturing may be the most intense way that firms interact with start-ups, it is not the only one. Another common approach is strategic alliances, when a large firm and a start-up agree to cooperatively pursue a project or product, typically in an emerging technology.

A key rationale for these joint efforts relates to the difficulty of assessing start-up firms. Financial investors often have a tough time figuring out whether a new biotechnology or semiconductor company really has a breakthrough technology. Young firms tend to turn to alliance financing when there is particularly great uncertainty about their prospects and those of their industry as a whole. Rather than trying to interest skeptical public investors or venture capitalists of their prospects, they turn to corporations for alliance funding. The larger company's greater insight into the nature of the technology the firm is working on allows it to make successful investments at times when uninformed public investors are scared away.[19] Like a venture deal, an alliance typically allows the larger firm to provide funding to the risky technology firm in stages, with payments being conditional upon, for instance, innovative progress, the production of a prototype, or the initiation of a phase II clinical trial.

Larger firms also attempt to address these concerns by entering into detailed agreements. Strategic alliance contracts can run many hundreds of pages, spelling out a wide variety of contingencies and how they will be addressed. The corporations are typically careful to reserve for themselves many of the key control rights, which allow them to guide the strategic directions of the projects. The time and effort devoted to negotiating these agreements appears to reflect the extent of the uncertainty surrounding the technology. In many cases, the larger firm will also take equity in the firm, in the hopes of making sure that everyone is "on the same page."

Strategic alliances have emerged as a primary way in which start-ups in many emerging industries get funded. But here, too, all is not simple. Numerous distortions can creep in, which lead to less-than-ideal contracts.

In many cases, the larger firms have ample financial reserves, while the start-up is needy. Particularly at times when it is hard to raise funds from the public markets, the smaller firm may feel compelled to undertake an alliance in order to survive. These imbalances can lead to the signing of one-sided agreements, where all the power is in the hands of the corporation, even though it would make sense for the start-up to make many of the decisions. Philippe Aghion and Jean Tirole developed a model that suggests that one-sided agreements will lead to fewer innovations than ones where neither of the parties suffers from financial constraints, and can arrive at an optimal agreement, which will typically be more balanced.[20]

In addition, the business development officers at the larger firms who negotiate these agreements face a dilemma. In many industries, the ultimate success or failure of the agreement will often not become clear for many years; in sectors like biotechnology, the verdict will not come in for a decade or longer. By this point, the negotiators will be off to a new job, probably at another company. Evaluating the business development group on the ultimate success of the deals they negotiate is almost impossible. Instead, firms tend to look at other things, such as the extent to which the terms in the agreement favor the corporation. These career concerns can also lead to agreements that are not the most conducive to success.

The dangers of the misshapen relationships that these pressures lead to can be illustrated by the alliance between Ciba-Geigy, a predecessor to the Swiss pharmaceutical giant Novartis, and young Silicon Valley drug delivery specialist ALZA.[21] ALZA had turned to an alliance strategy after its promising start-up ran aground: the California firm had quickly raised a substantial war chest from the public markets in the late 1960s and early 1970s, and just as quickly burned through it developing products that did not work. The two firms signed a research agreement in 1978, when ALZA was on the verge of

bankruptcy. Although the bulk of ALZA's technologies were covered
by the alliance, the young firm retained the right to also engage in a
variety of independent activities, including alliances to exploit technologies that did not conflict with the topics being jointly explored
with Ciba-Geigy.

Due to ALZA's financial weakness at the time of the alliance, Ciba-
Geigy was able to obtain vast control rights, such as eight of ALZA's
eleven board seats, majority voting control, extensive information
rights, and the ability to guide 90 percent of ALZA's research activities
through a number of review panels that were dominated by Ciba-Geigy
representatives. Despite these seemingly ironclad control rights, numerous tensions arose over the exact type of research the ALZA
researchers should be conducting.

In particular, Ciba-Geigy was concerned about other research projects and collaborations for which ALZA representatives kept seeking permission to establish with third parties. Although the boards
ultimately approved most of ALZA's requests, ALZA representatives
became frustrated at the long delays associated with the process. As
a result, ALZA scientists began bypassing the various review panels
and directly contacting senior Ciba-Geigy officials for permission to
engage in outside arrangements. While detailed reporting and monitoring processes had been stipulated in the original agreement, these
proved very difficult to enforce. Ciba-Geigy officials were also concerned that ALZA scientists were publishing material in journals that
disclosed their proprietary technology or might be employed in ALZA's
collaborations with other pharmaceutical firms.

As a result, Ciba-Geigy became increasingly reluctant to disclose
its own technologies in the area of drug delivery to ALZA. Ultimately,
these tensions led to the dissolution of the research collaboration at
the end of 1981. These conflicts, while perhaps extreme, illustrate what
can go wrong in these alliances. The one-sided nature of the agreement,

brought about by the differing positions of the two firms at the time it was negotiated, created a cascading series of problems that were ultimately impossible to resolve.

But just as with corporate venturing, in many cases, large and small firms *are* able to overcome their differences, and to create real value from their collaborations. Not only can we offer many stories of such fruitful partnerships between firms, but more systematic studies also document such success. For instance, Ben Gomes-Casseres, Adam Jaffe, and John Hagedoorn compare pairs of firms that form strategic alliances with otherwise similar firms that do not.[22] They seek to understand whether these pairs of firms successfully transfer innovative knowledge—a key objective behind the typical alliance. To measure this, they look at how frequently the firms cite each other's patents. While some knowledge flows are no doubt accidental or involuntary, if the alliances are fulfilling their intended role, greater interactions between the parties should lead to more mutual citations of each other's work.

They find that if firms have one recently formed alliance, there is only a modest increase (6 percent) in the probability that they will cite each other. As the relationship deepens, however, the knowledge flows accelerate: firms with alliances that are three or more years old have 19 percent more citations than nonallied firms, and pairs of firms with seven or more alliances have 63 percent more citations. These effects are particularly strong for firms that have a more intensive alliance—for instance, if one of the firms purchases equity in the other—or have headquarters close to each other. These patterns suggest that alliances, just like corporate venturing, do seem to be having real effects in boosting innovation.

Final Thoughts

Yes, many efforts to achieve a middle ground between the large corporation and the start-up have gone astray, whether due to a lack of clarity about a corporate venture program's mission or a poorly designed strategic alliance contract. But the large-scale evidence suggests that these programs have worked in creating value for the corporate parent and the young firm alike. The gains from unlocking the knowledge inherent in the large firm are sufficiently great to overcome a variety of limitations.

But while each model for boosting innovation can potentially learn from the other, naively mixing and matching—simply taking one feature from venture funds and one from corporate labs—is not a design for success. Thoughtfully designed hybrids can make a positive difference.

Before we turn to these implications, chapter 7 will consider another critical player: the public sector. Hybrid efforts like corporate venture programs, or innovation more generally, require some kind of government support, even if it's only by not getting in the way.

7

Leveling the Innovation Landscape

WELCOME TO MARSHALL, Texas, population 24,089.[1] Marshall, located near the Louisiana border in east Texas, along with nearby Tyler and Texarkana, has emerged in the past fifteen years as the patent litigation capital of the world. Yes, the city is the home of the annual Fire Ant Festival and the Stagecoach Days Festival, but is far from any major business center, much less any technology hub. Yet in 2008, 30 percent more patent cases were filed in east Texas than in the next busiest federal district (northern California).[2]

Marshall offers a combination that patent holders find irresistible: juries that are sympathetic and generous to plaintiffs and supportive judges. Fifty years ago, it was rare for a patent case to be heard by a jury anywhere in the United States: legal observers and practitioners alike felt that the combination of technological and legal complexity made these cases much more suitable for a judge to decide. Over the ensuing decades, however, patent holders discovered that juries were far more inclined to uphold patents and award large damages in these cases. Because the law assumes a patent is valid unless proven otherwise, juries in many cases seem to throw up their hands and assume it is valid, no matter how flawed the process for issuing the awards (defendants typically cannot describe how troubled the patent issuance system is during trials). For instance, Kimberly Moore, analyzing patent cases adjudicated in 1999 and 2000, found a huge disparity: juries were twice as likely to uphold patent holders as were judges.[3] Moreover, they tend to have a strong "home bias." Foreign patent holders who were pursuing U.S. infringers were far less likely to win than others: in fact, they were only about one-quarter as successful. Meanwhile, she showed that patentees based in the same state as the venue where the case was tried were disproportionately likely to succeed.

As a result, patent holders everywhere have turned to jury trials, and nowhere with as much gusto as in Marshall. The local juries' enthusiasm for patent holders may be reinforced by their knowledge that patent litigation has become the major economic development engine in town, as hotels and temporary offices fill up with free-spending litigators from Dallas, New York, and San Francisco. (For instance, in a recent survey, 93 percent of potential jurors in east Texas said that they favor protecting inventions and discoveries with patents, and 76 percent of these individuals said that they "strongly favor" patent protection.[4]) And with such a small population to work with, local lawyers can stock the juries with neighbors who are most likely to be sympathetic to the claims. Meanwhile, the judges of the U. S. District Court for the Eastern District of Texas have played an important role here as well. These solons are well known for putting these cases on a speedy time line with limited paperwork, creating a "rocket docket" where patent holders can be sure their day in court will come soon.

The result has been a series of monster verdicts coming out of the district courts. In 2009 alone, a jury there awarded $1.7 billion in damages against Abbott Labs (since sharply reduced on appeal), and a judge ruled that Microsoft had to go through the costly process of re-writing its Word program to eliminate the use of custom XML—which allows files to be smaller and more efficient—due to a problematic award to a small Canadian entity.

More generally, Marshall has become the global epicenter for liti-gation by what are termed nonpracticing entities—less kindly known as patent trolls. For instance, an entity known as American Video Graphics (AVG)—with a "coming soon" website and a Marshall phone number—purchased some video-game patents from equipment maker Tektronix. Included was an award (number 4,734,690), first filed for in 1984 and granted in 1988, that covers the 3-D view of the action in a video game. (Whether Tektronix really invented this feature, or was just the first to seek to get an illegitimate patent on an already-employed game feature past a harried examiner, is an open question.) Nearly two decades after the original patent filing, AVG launched a blizzard of litigation in Marshall against the video-game and device industry. Ultimately, most of the targeted firms chose to settle the dispute, fig-uring that the cost and risk of litigation were just too great. (Marshall juries have the reputation of being particularly unsympathetic to Asian companies as defendants, who dominate the video-game market.)

Repeated hundreds of thousands of times, this system becomes a substantial tax on innovation. Stories like this make it clear that there is some role for government in thinking about how innovation works—even if thoughts of government intervention recall Ronald Reagan's observation that "the nine most terrifying words in the English lan-guage are 'I'm from the government and I'm here to help.' "[5]

Can the government really boost innovation? Whatever one's po-litical stripes, it is indisputable that various branches of government do play a profound role, for better or worse, in influencing the effec-tiveness of firms in identifying and commercializing innovations.

Whether their activities focus on creating the backdrop for private sector activity or actively subsidizing innovation, the impact can be substantial. This chapter examines three ways that governments can influence innovation: making contracting clear, helping ensure that entrepreneurs have a shot at success, and priming the pump through direct investment.

Facilitating Contracts

If governments want to facilitate the creative approaches to promoting innovation, probably the most important step is to make it possible for firms to freely and confidently contract with each other. When we think about the interactions described in the last chapters—whether equity investments, strategic alliances, or investment partnerships— the critical importance of these arrangements is clear. If firms feel nervous about what these agreements really mean, or whether they can readily be enforced, the pace of collaboration—and of innovation itself—is likely to slow down substantially.

Numerous economists have suggested that the ability of entrepreneurs and investors to enter into complex contracts, with very different payouts if the company's progress varies, is highly beneficial. An example of such a contract would be convertible preferred stock, where the investor can choose either to get back the amount that he or she invested, or alternatively to convert the holdings into common stock. In cases where the firm does well, the investor gets all the upside of a shareholder. But if things get ugly, the investor has rights akin to that of a lender. These contracts are helpful, as they allow control over the firm to be transferred to the party that can make the best use of it. In particular, as discussed in chapter 4, these securities allocate control to the entrepreneur when things are going well, but allow the investors to assert control if the firm is doing poorly. In this way, entrepreneurs can be sure that if they do a good job running the firm, the investors will not be able to use their special rights to wrest away their hard-earned gains.

Antoinette Schoar and I examined how the contracts that venture capitalists and the firms in their portfolio enter into vary across developing countries.[6] In the analysis, we show that entrepreneurs and investors in countries with well-defined legal rules and effective court enforcement rely on these complex contracts, in which the assignment of control depends on the performance of the investment. In contrast, investors in countries with less well developed laws and courts are far less likely to use convertible preferred stock, and must instead rely on common stock, typically holding majority stakes in firms, which appears to limit their success. In the agreements involving preferred stock, the valuation the venture capitalists are willing to pay is sharply higher, or put another way, the slice of the equity the entrepreneurs must give up for a given amount of financing is considerably lower. But it is not just the entrepreneurs who benefit: funds that were active in developing nations where complex contracts are commonplace had an average return multiple 19 percent better than the returns of public stocks in that nation over the same time period, while those in socialist and civil law countries had a return 49 percent worse than the benchmark.

But, as the case of Marshall, Texas, shows, the legal framework has implications far beyond emerging markets. Patents—along with copyrights and trade secrets—are the key mechanism through which firms protect their ideas. Having a strong intellectual property portfolio is typically a prerequisite to interesting a larger partner into a strategic alliance. Similarly, if a start-up firm is seeking funding from a venture or angel investor, one of the first questions that potential backer will ask is about the barriers that will prevent others from imitating the idea.

Since the 1980s, patent policy in the United States—and to a lesser extent in other major industrialized nations—has increasingly been an obstacle to innovation, rather than its facilitator. Rather than promoting a system where there are well-defined rights to carefully delineated pieces of property, patent policy has become increasingly mired in uncertainty. Two changes, each not extraordinary in its own right, have combined to fundamentally alter patent practice over these years.[7]

First was the creation of a centralized appellate court for patent cases in 1982. Consistent with the history of specialized courts, over the last three decades this body has taken a consistently pro-patent tack. Viewed as a whole, its decisions have led to a broadening of patent protection, a lowering of the bar to prove infringement, an increase in the damages that patent holders can win, and an expansion of even more powerful weapons, such as pre- and posttrial injunctions, to be used against alleged infringers. Not surprisingly, patent holders have responded by filing many more suits (the number of suits filed more than quadrupled between 1981 and 2010).[8] Even the efforts of the U.S. Supreme Court, which has sharply questioned the federal circuit's logic in a number of important decisions such as *eBay v. MercExchange*, has not been able to reverse this trend.[9]

The second change was rising resource pressures within the U.S. Patent and Trademark Office. A combination of a many-fold increase in patent filings and fee diversions, wherein patent filing payments are transferred to the Treasury rather than being reserved for the patent office's operation, have put the office under severe financial strains. (Recent patent reform legislation did away with the fee diversions. Many observers anticipate, however, that the deficit pressures will lead to the reinstating of the diversions, due to the desperate hunt for revenues in the federal government.[10]) Low pay scales, high staff turnover, and poor relations between the employee union and management (with the consequence of stifling, difficult-to-eradicate work rules) have created a challenging work environment. Not surprisingly, the quality of patents being issued has varied dramatically across the office.

Together, these changes have created a stew that is increasingly toxic to young firms seeking to commercialize new ideas. Not only are firms filing and litigating more patents, but it is very unclear whether the bulk of rewards are going to the firms that are the true innovators. Instead, a whole army of firms has sprung up—with names like Acacia Research, Global Patent Holdings, and Intellectual Ventures—whose primary lines of business appears to be to buy up neglected patents, many of which have been issued with troublingly broad claims by inexperienced or harried patent examiners, and then sue others for extraordinary sums. (Larger companies, frequently ones who are losing out in the product market and seeking some way to maintain their profits, have gotten into the act as well.) Often they will threaten to shut down the entire operations of firms unless they can get a substantial settlement. And because the judicial system is so tilted to benefit the patent holder, alleged infringers frequently find it safer to settle than risk a catastrophic shutdown. These settlements inspire others to seek out yet more problematic patents.

The most dramatic illustration of the pathologies that can result is in the mobile phone industry. The enormous consumer demand for iPhones, BlackBerries, and Android devices has led to enormous stakes here, which have attracted patent lawyers like bees to a clover patch. David Drummond, Google's senior vice president and chief legal officer, estimated that a modern smartphone is covered by as many as 250,000 patent claims, many deeply problematic.[11] The magnitude of the problem can be seen from the experience of Research In Motion. The Canadian firm was essentially forced by the U.S. courts in 2006 to pay $612 million to patent troll NTP to prevent its BlackBerry devices from going dark. Many observers at the time expressed grave doubts about NTP's awards. Even the U.S. Patent and Trademark Office had indicated that it was likely to find all of the patents invalid in a reexamination proceeding, an indication the judiciary chose to ignore when forcing a settlement.[12] More recently, large companies have jumped into the mobile phone patent game, as seen by the $4.5 billion purchase of Nortel's patent portfolio by a consortium including Apple and Microsoft, and the Motorola Mobility acquisition by Google discussed in chapter 1. Whatever the merits of the patents being purchased, the fact that some of America's most innovative companies are spending many billions buying old patents from failed predecessors rather than using these funds to develop new ideas is deeply troubling.

Yet as problematic as the abuses of the patent system have been, Congress has addressed them with little success. The legislators spent much of the past decade considering a series of patent reform bills that, among other things, would have limited the ability of firms to file suits in districts where neither party has significant economic activity. These provisions have, not surprisingly, attracted bitter opposition from the well-heeled patent litigators and the licensing fee–engorged nonpracticing entities, both of whom have become adept at Washington lobbying. In a concerted effort to produce an "acceptable" bill that was ultimately enacted in 2011, the U.S. Senate stripped out most of the important provisions—including the rules that limited firms' ability to shop for a favorable judicial venue to file their case—from the bill that it passed.[13]

Sadly, litigation venue is just one manifestation of an intellectual property system that has gone seriously awry. We could discuss at length numerous other problematic features of the patent system, from the way that examiners are asked to adjudicate complex patent filings under intense time pressures and with little guidance to the lack of "apportionment" rules that can lead to huge damages for even minor acts of patent infringement. But this would fill an entire book—and then some. The copyright system's problems and their detrimental impact on innovation could fill another volume: illogically large statutory damages interact with huge legal ambiguities about what is legal and illegal, brought about by the fact that technology has moved much faster than the law. As surely as India's clogged civil courts deter foreign investors from investing nearly as much capital as the nation needs, the broken U.S. intellectual property system represents an impediment to effective contracting between innovative firms. Without confidence about who owns what, the willingness of investors to fund innovative firms and of large corporations to enter into creative transactions with them is substantially dampened.

Ensuring Entrepreneurial Health

The public sector also plays an important role in creating the environment for entrepreneurial companies to be successful *without* falling into the arms of large firms. Consider the experience of Japan. The nation undertook extensive efforts to stimulate venture capital during the 1990s and early 2000s, which proved to be remarkably ineffective in building long-run activity. A crucial barrier was that the environment in Japan was extremely unfavorable: from a labor market whose ground rules made it very difficult to hire talented managers to restrictions on the use of stock options for compensation, the system was stacked against the success of new ventures. Although entrepreneurs did turn to collaborations with established firms, one-sided deals that were negotiated out of desperation and often led nowhere were frequently the result.[14]

What does a short list of potential steps that can improve the health of entrepreneurial firms look like?

EASE THE COSTS OF FAILURE. One of the key aspects of innovative endeavors is the possibility of failure. Henry Ford's misadventures with his Detroit Automobile Company and Henry Ford Company before beginning Ford Motor Company are a paradigmatic example. Failure is often seen as a necessary rite of passage in Silicon Valley, to the point where there is today an annual conference for technology entrepreneurs, FailCon, to celebrate and learn from failed ventures.[15] The contrast with the attitudes that have characterized most of human history could not be sharper. The notion of debt bondage—that borrowers can offer themselves as security, and become slaves of their lender if they cannot repay their obligations—was already enshrined in the Code of Hammurabi in the eighteenth century BC and remains commonplace in much of the developing world today.[16] The more "civilized" alternative of the debtors' prison has persisted in many nations into the twenty-first century, and even in de facto form in the United States.[17]

These rules appear to have very real consequences. Viral Acharya and Krishnamurthy Subramanian show that firms in technology-intensive industries patent more, and those patents are more frequently cited, if they are based in nations with forgiving bankruptcy laws than in countries that are pro-creditor.[18] Similarly, they show that countries that switch from pro-creditor to pro-lender bankruptcy policies experience a flowering of innovation, at least in part because firms are willing to take on debt to finance the development of their new ideas. More generally, Augustin Landier shows how a society that punishes failure severely can create a vicious cycle, with entrepreneurs sticking with struggling firms lest they be regarded as failures themselves, which in turn leads to even more sanctions for those who do fail.[19]

ENHANCE THE POOL OF TALENT THROUGH IMMIGRATION.

A huge share of advanced students in science and technology today in the United States—and many other industrialized nations—are immigrants. For instance, in 2008, 48 percent of all those earning science and engineering doctoral degrees were temporary residents, as were nearly 60 percent of all postdoctoral students in these fields.[20] (These statistics understate the impact of immigrants, as they do not count green card holders and immigrant citizens.) These individuals do not just excel within the academy: one study found that there was at least one immigrant key founder in over a quarter of the technology companies established in the United States between 1995 and 2005.[21]

Yet many of these individuals have struggled to remain in the United States after completing their studies, wrestling with onerous restrictions associated with getting and retaining green cards, H1-B visas, and the like. Again, the U.S. Congress seems incapable of addressing the issue. A dramatic contrast is Start-Up Chile, a program the Latin American government began in the fall of 2010.[22] It tempts foreign entrepreneurs with a stipend of $40,000 a year, a one-year residency visa, and a dedicated team of seven people to provide guidance in navigating the business culture of the country. This effort was part of a pledge by President Sebastián Piñera to add one hundred thousand new businesses to the Chilean economy by 2014, which, he argued, would require the nation to look outside its borders "to regain its entrepreneurial and innovative culture."[23]

FACILITATE LABOR FLOWS ACROSS FIRMS. One of the defining characteristics that allowed the emergence of Silicon Valley was the fact that California did not allow the enforcement of noncompetition agreements. Unlike in most states, employees here could not be forced to sign an agreement committing themselves not to work for a competitor for several years after leaving the firm. The rationale for such agreements—that it allows the employer to invest more in a worker, secure in the knowledge that he or she is not going to run off to sell his newfound skills to someone else—is an old one in English common law, and adopted by most U.S. states. But as Ron Gilson argues, California's unique history, and its legislators' struggles to develop a legal code that blended its Spanish, Mexican, and English legal traditions, led it to eschew these agreements.[24] "The existence of this anachronistic legal rule during Silicon Valley's development," he argues, was crucial in shaping the entrepreneurial culture that we see to this day.

These arguments are borne out by Matt Marx and coauthors, who explored the consequences of Michigan's "accidental" switch from prohibiting to allowing noncompetition agreements.[25] (The provision was included as part of a general legal reform, apparently without the awareness of the legislators.) Not surprisingly, once this change was enacted, the mobility of Michigan workers relative to their peers in other states dropped sharply, by over 8 percent. Also expected was the fact that the drop was considerably more dramatic for workers with highly specialized skills. What is more striking was the impact of the legal change on the movement of inventors. After the policy shift, there was an increase in the number of inventors who left Michigan for states that do not allow noncompetition agreements (again, relative to those in other states). One might wonder whether this effect is driven by the automobile industry's turmoil, but the pattern still holds after eliminating those working on auto-related technologies. The brain drain is particularly pronounced among the most productive inventors.

MAKE THE MOST OF THE ACADEMIC BASE. Many nations have invested huge amounts building up their ability to do academic research. But the returns from these efforts in terms of economic activity have often been modest. In many cases, at least some of the blame must be laid on underinvestment in technology transfer offices, which serve as key conduits between the academic laboratory and the market. An underresourced office consisting of junior people looking to escape their poorly paid positions is not a road map to success. Inappropriate incentives for these offices are another challenge: often the offices are pressured to generate the most up-front revenues they can. For instance, at one of the national laboratories Adam Jaffe and I visited, researchers chafed under the technology transfer offices' demands for payments and royalties that they considered excessive, given the early stage of the technologies.[26] Some felt that the incentives offered the officers led them to seek to license technologies to large corporations that could offer larger up-front payments than start-ups, but which seemed less likely to nurture and ultimately bring the technologies to the marketplace.

RECOGNIZE THAT TAX POLICY HAS AN IMPORTANT ROLE TO PLAY. In so many nations, discussions about taxes have taken on an almost quasi-religious tone. Far be it from us to wander too deeply into these treacherous waters, but a line of research dating back to a 1987 paper by Jim Poterba is worth noting.[27] Since then, economists have widely accepted the proposition that decreases in capital gains tax rates should increase the attractiveness of becoming an entrepreneur. Poterba argued that increasing the differential between the tax rates on capital gains and ordinary income was critical: such a gap would spur corporate employees to found companies.

Most entrepreneurs, even those running high-potential, high-risk ventures, are not fresh-faced kids who have just completed (or dropped out of) school, despite what the celebrations of Sergey Brin, Mark Zuckerberg, and their cohort would suggest. If the typical would-be entrepreneur faces substantial tax burdens from his or her gains from a new venture, he or she will be less likely to leap over the fence. Creating such incentives need not entail cutting all capital gains taxes. One approach that has been employed in many countries is to create special tax rates for capital gains from investments in entrepreneurial firms. For instance, in the United Kingdom, to improve the fiscal environment for entrepreneurs and their investors, effective capital gains tax rates on the disposal of business assets held for more than two years have been reduced from 40 percent to 10 percent.[28] Given the evidence on the effectiveness of capital gains tax cuts and the very real revenue needs that many governments face, such targeted measures represent an attractive middle road.

EASE THE PROCESS OF GOING PUBLIC. Having a viable avenue to exiting investments through an initial public offering has two advantages: it allows firms and investors to gain an attractive return on their investments and, for the majority of new ventures that do not go public, the possibility of such an offering provides a validation of the firm's value. Yet in many developed nations, regulatory changes after the collapse of the dot-com bubble and scandals such as Enron and WorldCom—while typically motivated by concerns about the behavior of much larger firms—have raised the barriers to these offerings. Nowhere are these effects more dramatic than in the United States, where the share of IPOs of worldwide entrepreneurial firms has dropped sharply in the past decade.[29] In 1999, over half of the world's offerings took place here; this had fallen to 10 percent in 2009, although there has been a modest recovery since.

Among the policy steps put in place during the interim were the 2002 Sarbanes-Oxley governance reforms. While the abuses that the reforms sought to address were reprehensible, the costs of compliance have taken a heavy toll on smaller, entrepreneurial firms. Peter Iliev documents the impact of the Sarbanes-Oxley Act by employing a feature of the law: small firms, defined as those with the lowest value of shares traded, received a "stay of execution" from complying with the law's substantial requirements.[30] By comparing firms just above and just below this threshold, he shows that the impact of the act was substantial. The small firms that needed to comply with Sarbanes-Oxley nearly doubled their annual audit fees. This increase of almost

$700,000 was quite significant for these firms. During the two years between the announcement of the rule and the filing of the first annual reports under the new law, the firms that needed to file had 17 percent lower stock returns than their peers just below the cutoff. This pattern suggests that the market saw the cost of complying with this new rule as destroying a substantial fraction of these firms' value. While the JOBS Act, enacted in the spring of 2012, may ease these challenges in the United States, more remains to be done.

Perhaps surprisingly, Russia has been the country working most vigorously to address these barriers.[31] In September 2009, President Dmitry Medvedev bluntly asserted that dramatic reform would be absolutely essential to position the nation for the twenty-first century. His concerns were borne out by Russia's poor rankings on many measures of innovation friendliness: for instance, in 2009 the country placed 45th out of 70 ranked countries in the technical skills of its workforce, and 101st out of 115 countries in intellectual property rights protection. Medvedev argued that the Russian government could act as a facilitator, rather than a barrier, in Russia's attempt to emerge as a global innovator: "Public and private companies will receive full support in all endeavors that create a demand for innovative products. Foreign companies and research organizations will be offered the most favorable conditions for establishing research and design centers in Russia. We will hire the best scientists and engineers from around the world."[32]

Many of the initiatives that he has promoted are focused on the Skolkovo innovation hub. An estimated $500 million of public and private investment will go to building an "innovation city" that will mimic California's Silicon Valley. To do this, the government rolled out a wide variety of incentives. These include zeroing out profit, property, and land taxes and sharply reducing required payments for social insurance programs. Selected companies would also receive re-imbursement of customs payments, expedited visas for foreign work-ers, state investments, and permission to sell to the government using streamlined procurement procedures. To address concerns about Rus-sia's overworked and frequently corrupt judicial system, the plan calls for an antipiracy court in Skolkovo, with experts in intellectual prop-erty dealing with complex cases.

The jury is still out on how effective this effort will be. Certainly, success is far from certain: many other ambitious efforts promoted by bold leaders have fallen prey to poor design, incompetent implementa-tion, and outright corruption. And as many companies and investors have discovered to their chagrin, Russia can be a particularly challeng-ing business environment. But the Skolkovo effort stands out for its breadth of response to these frequently encountered barriers.

Priming the Pump

The public funding of innovation was established as a centerpiece of U.S. science policy with the release of Vannevar Bush's *Science, The Endless Frontier* in 1945.[33] And from its inception, postwar science policy relied on private sector firms—typically large contractors—to pursue many of the innovations that the government sought in arenas such as defense and space. Later on, entrepreneurial firms were added to the federal innovation funding mix, with the initiation of the Small Business Investment Company program in 1958 and the Small Busi-ness Innovation Research program in 1982. In a more recent example, the European Commission wholeheartedly endorsed the public fund-ing of private innovation with the adoption of the Lisbon Agenda in 2000, which sought to increase R&D expenditures as a share of GDP from an average 1.8 percent of GDP in the late 1990s to 3 percent by 2010, at least one-third of which was to be funded by governments.[34]

The appeal of these policies is easy to understand. Economists have long argued that innovation is a "public good": much of the knowledge from R&D "spills over," or finds its way through various channels, to competitors. As a result, a typical firm is likely to do less research than society as a whole would desire. Moreover, as discussed in Chapter 1, there is a well-established link between innovation and economic growth. To ensure the desired amount of innovative activities, public subsidies—whether in the form of outright grants, tax credits, or more newfangled structures such as "advanced purchase commitments"—should be able to play an important positive role.

But despite this compelling logic, policies to promote innovation have frequently led to unexpected and far less gratifying consequences. A lengthy saga of horror stories could be told as to how government policies to promote innovation—whether by established or entrepreneurial firms—have gone astray. In contrast, the efforts that have been the most successful have shared certain elements.

The first of these has been letting the market provide direction by linking funding to the ability to provide or raise matching funds. By insisting on these complementary funds, policy makers can limit the danger that public monies go to appealing but impractical projects, whether an initiative around a technology in which the nation has no real expertise or an effort to build an innovation hub in a remote and ill-suited region.

Doubtless the greatest illustration of the power of matching funds must be the Israeli Yozma program, which in the early and mid-1990s sought to develop Israel's nascent venture industry.[35] As a result of the program, the Tel Aviv area has emerged as one of the great venture hubs worldwide, in recent quarters typically ranked fourth in compilations of venture activity by metropolitan region, such as PriceWaterhouse-Coopers' "Money Tree Report".[36] One decade after the program's inception, the ten original venture capital groups funded by the Yozma program were managing funds totaling $2.9 billion, and the Israeli venture market overall had expanded to include sixty groups managing approximately $10 billion.[37] Central to this program was the provision of matching funds to venture groups that were willing to commit to investing in innovative Israeli companies (with added "sweeteners" if the venture group was particularly successful). The total public expenditure of the venture capital program was about $80 million, which was repaid with interest by the funds.

A second guidepost is to ensure the program is governed by straightforward and transparent rules. Many public efforts have become so burdened with requirements that the most promising innovators end up shying away, concluding the programs are more trouble than they are worth. As a result, many of the funds may end up with lower-tier firms, whose ability to boost innovation may be considerably less. In other instances, the decisions regarding funding are opaque. Such ambiguities can open the door for problematic behavior of various types, from government officials channeling funds to friends and family to intensive lobbying by private sector entities to get selected. Defining and adhering to clear strategies and procedures for funding and creating a firewall between elected officials and program administrators can help limit such problematic behavior.

The Advanced Technology Program, which provided funding for innovations at small and large U.S. firms alike between 1990 and 2007, provides an illustration of the first point above: the costs of excessive complexity.[38] The program managers defined a complex set of criteria for evaluating programs, including requirements that firms be working on precommercial research and strong preferences for proposals between consortia of firms. But the program managers were less adept at discerning whether the applying companies were viable vehicles for actually accomplishing their innovation goals. Reflecting in part the complexity of the evaluation process, a distressingly large number of awards ended up in the hands of companies that had already accumulated numerous government grants. These entities often had a "contract research" mentality, and stayed in business largely on the basis of these awards. Such companies in many cases seemed to avoid accountability indefinitely and over time became experts in the grant application process itself.

A third marker is to ensure that the efforts are appropriately sized. On the one hand, too small an effort may do little besides inflate expectations, leading to unhappiness when few results are apparent. Too-large programs, on the other hand, can create a wide variety of distortions, as public funds crowd out private investments.

An effort that went against this principle—as well as the second one—was the cleantech initiative that was part of the controversial stimulus bill program that the Obama administration championed in 2009.[39] The funding of cleantech innovation was an important element of the American Recovery and Reinvestment Act of 2009. But the program was not well thought out. The awards intended for entrepreneurial firms were very substantial, totaling at least a few billion dollars and equaling or exceeding the venture sector's investment in cleantech in 2009 (a total of $1.9 billion).[40] Moreover, the award process was characterized by a lack of clarity. In this uncertain environment, it is not surprising that venture capitalists and entrepreneurs responded in two ways. First, many venture groups appeared to hold back making new investments until the recipients of the federal largesse were clear, a process that took many months. Second, a number of firms sought funds the good old-fashioned Washington way: by hiring lobbyists. A partner at leading venture capitalist New Enterprise Associates suggested that at least half of the twenty-five cleantech firms in its portfolio hired Washington representatives. And these expenditures seemed to pay off. For instance, the big winner of the Department of Energy's battery funding orgy, A123 Systems (with $249 million in awards), spent about $1 million on Washington representatives between 2007 and early 2009.

A fourth principle is the need to view public investments in innovation as long-term ones. To expect quick returns from investments in innovative activities by either large or small firms—as many advocates seemed to suggest when advocating the 2009 stimulus act—is naïve. As an extensive literature by sociologists, economists, and historians of technology suggests, the diffusion of new inventions is frequently a prolonged process, where even ideas that seem obvious in hindsight (e.g., the Internet) take a long time to catch on. Yet often impatient policy makers or legislators seek rapid returns from their investments, which can defeat the goal of creating innovations in the first place.

Consider, for instance, the shifts over the past decade at the Defense Advanced Research Projects Agency (DARPA) of the U.S. Department of Defense.[41] DARPA has long been famous for making bets on researchers working on emerging technologies at their earliest stages, such as when in 1966 Charles Herzfeld after an hour's discussion (and no formal proposal) agreed to fund the development of what would become the Internet. Many of today's critical information technologies, from the computer routers commercialized by Cisco to the search algorithms underlying Google, emerged from DARPA-sponsored research. Yet as the President's Information Technology Advisory Commission and others have highlighted, DARPA, over the course of the George W. Bush administration, moved away from basic academic research, putting a far greater emphasis on short-term projects with well-defined deliverables. Even in areas as seemingly fundamental to national defense as cybersecurity, the agency has abandoned funding projects likely to take more than five years to come to fruition. By bringing more emphasis on short-term deliverables and time lines, DARPA's managers doubtless see themselves as getting more "bang for the buck." But the effect may be precisely the opposite.

A final guidepost is to undertake careful evaluations of initiatives—and then to act on these evaluations. Few programs are perfect as they are initially conceptualized. Yet all too often, decisions about whether to continue programs or not end up being driven by a few high-profile examples. Even worse, many programs continue with no careful examination, recalling another classic Reagan line, that "a government bureau is the nearest thing to eternal life we'll ever see on this earth."[42] It is not surprising that many of the nations with the most successful technology policies, such as Israel and Singapore, have shown a much greater willingness to review programs seriously, and then to cut off those that are underperforming.

One illustration of the potential power of such examinations is the experience with the research and exploration tax credit in the United States.[43] Introduced in the Economic Recovery Tax Act of 1981, it has been a primary means of subsidizing research over the past three decades. But hardly had the bill been enacted than economists began pointing out a design flaw in the legislation: the tax credit was computed based on incremental R&D, or more specifically, on spending that exceeded the average level in the past three years. This provision, which rewarded firms who altered their R&D expenditures sharply from year to year, ran counter to what we know about what makes a long-term research agenda effective. Second, such a provision may tempt firms to delay increasing their R&D expenditures in a given year, in order to keep the base lower in future years—exactly the opposite of what policy makers intended.

Ultimately, these evaluations had the desired impact. Congress heeded these complaints and changed these provisions when renewing the credit in 1989. In particular, the way that the tax man identified extra research was altered to use information on R&D spending from the more distant past.[44] In this way, the temptation for firms to "game" their taxes by manipulating their R&D spending ratio from year to year was greatly reduced.

Final Thoughts

The challenge for public policy is to keep the economic environment well tuned so that entrepreneurship and contracting between firms can flourish. Direct interventions must be approached cautiously and policy makers must proceed with an eye to catalyzing private sector activity rather than creating new activities out of whole cloth. But these principles, however simple they sound, prove difficult to execute. But the success of some of the best-designed initiatives suggests the power of thoughtful interventions.

In the final chapter, we turn back to the private sector, considering the ways that hybrids between the corporate laboratory and the start-up can prove most effective, highlighting the abundant potential for cross-fertilization and learning.

8

Improving the Design

FTER WORKING THROUGH both corporate and venture-funded R&D, we could reasonably conclude that the two areas are each ideal in their own way. Although each has its weaknesses and drawbacks, each has also demonstrated successes; their continuing roles in the economy cannot help but impress. While neither model is perfect, they seem better than the alternatives.

But there are some powerful reasons for thinking that other alternatives might exist. The past several decades have seen numerous changes in the nature of emerging technologies and the ways that innovation is pursued, mostly associated with the widespread dissemination of the Internet and the explosion in computing power. This fact alone might lead us to wonder whether institutions developed in the nineteenth or mid-twentieth centuries are still the right answer. We have seen both corporate research executives and venture capitalists expressing dissatisfaction with the limitations of their models. As chapters 3 and 5 discussed, both sectors have seen substantial experimentation and change over the past decade. Despite the strong inclination of many senior venture investors to be confident of the perfection of the traditional model, the evaluations of hybrids between corporate and venture innovation discussed in chapter 6 suggest they have real promise.

In this final chapter, we seek to understand where the ideal territory between these two well-established institutions may lie. We will highlight a number of ideas as to how each might more effectively generate innovation. In some cases, these two institutions can learn from each other: there are lessons of the corporate research laboratory for the venture capital sector, and vice versa. In other cases, they can simply address design flaws that have sprung up because of history or accident.

A caveat about this chapter is in order. This book, as highlighted in the preface, is dealing with a territory where much is still unknown. I have sought to marshal what evidence there is—whether theoretical works, case studies, or large-sample analyses—to support the various arguments. This last chapter is rather different, in that it represents some informed guesses as to how the industry could change for the better, and where these shifts might take us. As a result, all the initial caveats apply tenfold.

Lessons for Venture Capital[1]

Looking at the venture model, it is easy to reject the all-too-frequent claims that the sector is fundamentally broken or dying. This is not to say that there are not ways to make this more fecund as a source of funding for innovation.

The first such suggestion relates to the duration of venture partnerships. Since the early days, these funds have been structured for an eight- to ten-year lifespan, with provisions for one or more one- to two-year extensions. Venture capitalists typically have five years in which to invest the capital, and then are expected to use the remaining period to harvest their investments.[2]

The uniformity of these rules is puzzling. Funds differ tremendously in their investment foci: from quick-hit social media businesses to long-gestating biotechnology projects. In periods when the public markets are enthusiastic, venture capitalists may be able to exit still-immature firms that have yet to show profits and, in some cases, even revenues. But as discussed earlier, there has been tremendous variation in the public investors' appetite for such firms. It is not surprising that the venture funds have increasingly focused on sectors such as software and social networking, which are characterized by fast innovation "clock speeds."

Certainly, within corporate research laboratories, great diversity across industries exists in terms of the typical project length. What explains the constancy of the venture fund lives?

One possible explanation is that a reasonably short fund life seems to have been the norm in limited partnerships of all types. The structure of venture capital limited partnerships seem to have been largely borrowed from pools devoted to other endeavors. Many of the other arenas where limited partnerships were employed in the twentieth century, such as real estate, oil and gas exploration, and maritime shipping, were reasonably short-lived. In the formative days of venture partnerships, the lawyers drafting the agreements may thus have gravitated to relatively short fund lives. With the passage of time, such arrangements have been taken as gospel by limited and general partners alike.

Another factor behind the persistence of the ten-year agreement has been resistance of limited partners to long fund lives. Investors may fear that if they give the funds to a substandard venture group for a longer period, they will be stuck paying fees until the end of time for very limited returns. This reluctance may tell us more about the outsized nature of the fees that venture funds receive than about the inherent desirability of a uniform-lived fund.

One natural reform would be to better link venture fund life to the nature of the investments being pursued. In this way, funds addressing more challenging areas, where profitable firms and successful exits are hard to achieve, could have longer to work with their firms. Such flexibility would limit the likelihood that venture funds would be driven by excessively short-run considerations, while still maintaining the critical protection that investors enjoy in limited partnerships: the fact the groups cannot hold on to their funds forever, but must ultimately return the capital to their investors and persuade them to reinvest or others to contribute capital.

To address the concerns of institutional and individual investors about long fund lives, a natural step would be to ease the process through which funds can be wrapped up. As it now stands, it is relatively easy for investors to dissolve a fund where the manager does something outrageous, such as being convicted of embezzlement or insider trading. But winding up a partnership for poor performance is much harder, since it is much more difficult to prove in court. The primary mechanism by which limited partners can practically terminate an underperforming fund is through a so-called "no-fault divorce" clause, which allows the investors to simply dissolve the fund without a finding of any misbehavior. But today, many partnerships either do not include such a clause or make it conditional on the approval of a very large supermajority of the limited partners. A reasonable route would be to allow longer-lived funds for venture capitalists operating in slow-gestating technologies, but lower the hurdle required for a no-fault divorce in the latter years of the funds' lives.

A second recommendation is to better link performance to the compensation of venture capital groups. This point might initially seem puzzling since we held up the start-up model, with its strong emphasis on equity holdings, as something that corporate research labs might do well to consider when rewarding researchers. But even though the linkage between rewards and outcome is far better for the entrepreneurs than for corporate scientists, when it comes to the venture investors themselves, distortions have crept in during recent decades.

How venture capitalists are compensated has changed little, even as the funds have grown much larger. Venture groups usually receive a share of the capital gains they generate (typically 20 percent, but sometimes as high as 30 percent) and then an annual management fee (often between 1.5 and 2.5 percent of capital under management, though it frequently scales down in later years). Such a fee is quite modest for a fund of a few million dollars: it is likely to cover only a very modest salary for the partners once the costs of an office, travel, and support staff are factored in.

But this compensation structure has remained largely unchanged as funds have become substantially larger. And as venture groups begin managing hundreds of millions or billions of dollars, substantial "economies of scale" appear: put another way, as a group becomes ten times larger, expenses increase much less than tenfold. As a result, management fees themselves become a profit center for the firm. And these steady profits may create incentives of their own, not all of them appealing: for instance, the temptation to raise a larger fund at the expense of lower returns, the lure of doing a subpar deal so money can be put to work quickly and a new fund can be raised sooner, and a tendency to do excessively safe investments that will not have as much upside, but pose less possibility of a franchise-damaging visible failure.

Just how large a temptation the venture capital compensation scheme can pose is illustrated in table 8-1. The ninety-four venture funds (the partnerships were raised between 1993 and 2006) tended to be among the larger groups in the industry, with an average size of $225 million. The table presents the compensation for the average fund in their sample, with the payments computed both per $100 raised and per partner. In each case, the sum of fees and carried interest

TABLE 8-1

Total compensation of venture capitalists

Present value of	Mean
Carried interest per $100	$8.36
Management fees per $100	$14.80
Total revenue per $100	$23.16
Carried interest per partner (US$ M)	$6.55
Management fees per partner (US$ M)	$10.57
Total revenue per partner (US$ M)	$17.11

Source: This tabulation is based on the work of Andrew Metrick and Ayako Yasuda, who worked with a large intermediary that obtained detailed organizational information on venture capital groups in which they invested or that were seriously considered. "The Economics of Private Equity Funds," *Review of Financial Studies* 23 (2010): 2303–2341.

collected by the fund managers are reported. Because these payments are received over time, they are expressed in dollars normalized to the time of the funds' closing.[3]

Two striking patterns emerge. The first is simply how large the compensation is: of every $100 invested by the limited partners, over $23 ends up in the pockets of the venture investors. The sum that each partner receives is substantial: essentially, on the day that the average fund closed, each partner earned the equivalent of $17 million.[4] These sums are particularly striking when one considers that many groups were raising funds every three or four years.

These sums might not be disturbing if they were being driven by incentive compensation: if the very substantial payouts to each partner reflected the large amount of money being made by the limited partners in the fund. But profit sharing is not the most important source of compensation. Instead, almost two-thirds of the income is coming from the management fees, which remain fixed whether the fund does well or poorly.

In all fairness, this pattern is not just a consequence of greedy venture capitalists. Limited partners frequently look the other way, or tacitly encourage such fee structures. An illustrative anecdote was a venture capitalist who decided to head out on his own after a number of successful deals at a larger fund. Frustrated at the inappropriate incentives in the industry, he proposed to raise a fund with a significantly lower management fee but a higher profit share than the industry norm. The almost universal reaction among limited partners was that they were impressed by his accomplishments, but would be much more enthusiastic if he would return to the market with a traditionally structured fund. When he pushed on this point with the private equity staffers at a pension where he had close ties, he was told that the deal terms would lead to awkward conversations with the investment committee whom they reported to: once they understood the advantages of his proposed structure, they would start demanding it of all funds. The investment staff did not want to hazard the fallout that would ensue.

What explains the traditional reluctance of individuals running the private equity groups at the pension funds and other institutions to push to change these compensation arrangements? Staff members may not really understand the economics of the funds, they may fear that rocking the boat would limit their own ability to get a high-paying position at a fund or an intermediary in the future, or the officers may worry that developing a reputation as an activist would jeopardize their organization's ability to access the best funds. The last concern is a reasonable one. After the giant California Public Employees' Retirement System led a consortium of pension funds that pushed for an overhaul of private equity compensation in the mid-1990s, they were shunned by venture and buyout funds alike.[5] In recent years, we have seen more collective discussion of these issues by limited partners in meetings of the Institutional Limited Partners Association. But many of their proposals have been modest half-measures to adjust fees without addressing the more fundamental issues.

What makes this state of affairs particularly frustrating is that an alternative model beckons: the way that venture capital groups used to operate. Early venture capital groups, such as Draper, Gaither & Anderson, negotiated budgets annually with their investors. The venture capitalists would lay out the projected expenses and salaries, and reach a mutual agreement with the limited partners about these costs. The fees were intended to cover these costs, but no more. (A few "old school" groups such as Greylock still use such an arrangement.) Such negotiated fees greatly reduce the temptation to grow at the expense of performance and are likely to ultimately lead to more successful and innovative start-ups.

A third recommendation relates to dampening the distortions of the fund-raising process. The need to impress investors around the time of raising a new fund can make venture capitalists do some ugly things. Such actions may be beneficial to their bank account balances, but certainly do not enhance the prospects of the innovative companies they fund. One way to address this problem is to move away from the current model, where groups raise a single fund every few years. In this setting, where a single fund-raising cycle will shape the prospects of the group for years to come, the pressures to game the system may just be too great.

One example of a group that has taken this remedy to heart is General Atlantic.[6] Founded in 1980, the firm initially invested solely on behalf of Charles Feeney, who had grown rich building up the duty-free company DFS Group. The firm today has $15 billion under management, and invests in a wide range of deals, from venture capital to traditional buyouts. Among American groups, it was one of the early investors overseas, and today has offices in London, Düsseldorf, Hong Kong, Beijing, São Paulo, and Mumbai.

One of the unique aspects of General Atlantic is how its fund-raising works. Rather than raising a single fund every three to five years, as is standard, the organization simultaneously operates multiple funds on an annual basis. Essentially, a number of limited partners commit each year to a fund. It might be thought that this system would make only the lawyers happy, who doubtless keep busy drafting agreements each year. But this feature also allows General Atlantic to avoid the fund-raising cycle that dominates the thinking of so many groups. Observes managing director John Bernstein, "We don't have a typical fund structure, where you go out, raise a fund and invest for a period of time. Each investor has their own fund, with different investors renewing at different times, which means we end up with greater stability. This is different from the wider industry, where fundraising has put cyclicality into the structure."[7]

This intriguing model should be considered by other groups, and demanded by limited partners. Because fund-raising is so ubiquitous in this setting, the temptation to undertake steps that sacrifice the best long-term interests of the firm in order to impress the limited partners in the short run is greatly reduced.

A related point has to do with the honesty of the numbers presented in the fund-raising process itself. Anecdotally at least, it appears that that a lot of "creativity" characterizes the way in which venture groups present their past performance, with the most license not surprisingly being taken by the most marginal groups. These observations suggest the desirability of a more systematic approach to calculating performance, whether the valuation of transactions still in the venture capitalists' portfolios or the computation of the rate of return. One approach would be to create an independent, industry-funded intermediary that could certify that groups employed a standardized approach. Such consistency would permit more ready comparisons across groups, and help ensure that better investment decisions are made.

Some might object that the proposed changes are too hard for any individual venture capitalist or investor to accomplish alone. Rather, they would entail changing the attitudes of a large number of parties, to arrive at a new industry equilibrium. However reasonable these concerns, we should not forget the power of experiments that can be undertaken by a smaller number of actors, and might pave the way for broader shifts. For instance, in November 2011, the Texas Teachers' Retirement System created two dedicated funds in collaboration with Apollo Global Management and Kohlberg Kravis Roberts, which are envisioned to run for fifteen to twenty years. While these two funds will invest primarily in later-stage transactions, such as buyouts and distressed debt, they illustrate the broader point: that a creative structure does not require a fundamental reworking of an industry. In this case, a handful of visionaries at a single pension fund, including chief investment officer Britt Harris, senior managing director Steve Le-Blanc, and senior director Rich Hall, envisioned and implemented this sharp departure from industry practice.

Lessons for Corporate R&D

The first set of lessons for corporate R&D—about the importance of commitment, compensation, and the tolerance of failure—can be drawn from looking at the elements that have made venture firms successful. The last two, relating to the need for knowledge transfer and experimentation, address needs specific to the corporate context.

Before we plunge into the specifics, it is worth recalling the old hippie adage: "Think globally, act locally." The challenges discussed in this volume affect the largest research-performing institutions around the globe. But many of the changes outlined below can be made in a single firm, or even a single laboratory or new product group. In fact, as I will emphasize, it is probably for the best that the changes be approached on such a piecemeal basis.

One of the key lessons from the venture sector has to do with *staying power*. The commitment that an institutional investor makes to a venture fund is a binding one: even if the limited partner contributes a small amount of the total capital promised at the time of closing, there is an expectation that the total sum promised will be provided. Even during the depths of the financial crisis, it was rare for investors to walk away from these commitments (a step that would have led to various sanctions, such as the forfeiture of the amount invested to date, as well as a damaged reputation). Instead, limited partners tended to sell stakes in these funds to other investors who were more liquid. In some cases, the limited partners even **paid** other groups to assume their commitments.

Now contrast this experience with that of corporate venturing funds. Companies have been all too fickle in their commitment to corporate venturing. Often, simply the accession of a new senior officer—a replacement CEO, chief financial officer, or R&D head—has been enough to trigger the abandonment of earlier efforts: it is almost a corporate ritual to discard the pet projects of one's predecessor!

This historical lack of commitment has, of course, real consequences. Employees seem less likely to join a corporate venturing group, entrepreneurs are reluctant to accept their funds, independent venture funds are hesitant to syndicate investments with these groups, and corporate-funded start-ups find collaborations harder to arrange. In each case, the very real possibility that the rug will be pulled out from under the corporate venture initiative leads others to be reluctant to work alongside them.

Corporations can benefit from the process of formal commitments. Several steps can make these programs costly and time consuming to unwind: entering into legal agreements governing the fund and its economics, putting funds for these efforts into escrow, involving third parties in fund management, and incorporating the portfolio companies as independent legal entities with outside shareholders.

This advice might seem counterintuitive to many. Isn't flexibility a good thing? Why make it difficult to change strategy? Moreover, in some cases, opportunistic entrepreneurs have ended up suing their corporate venture backers when their ventures failed, even if the failure simply reflected the technological challenges encountered along the way. These suits often seem motivated by a cynical calculation that the deep-pocketed corporation would rather settle than undergo a bruising and uncertain fight. Commitments like the ones recommended here might leave the corporation more vulnerable to such actions. But despite these costs, the benefits from these steps are likely to be substantial. By highlighting their commitment to these programs, corporations can help address the concerns engendered by the faddish behavior in the past, and attract a higher caliber of partner to the effort.

A second area where corporations should borrow ideas from the world of venture funds and prize competitions relates to *the design of compensation schemes*. Corporations face a major challenge when shaping reward structures for those who undertake and supervise innovative activities. Traditional rewards that link payments to short-run performance may lead to an unwillingness to take chances. Ultimately, given the difficulty of supervising innovative projects, having the power of long-term incentives seems essential. Moreover, in an era like today's, where high-powered incentives for innovators are increasingly available elsewhere, whether from start-ups or prize competitions, an organization that does not hold out the possibility of substantial rewards will likely lose its best innovators.

A closely related point where the world of venture capital should provide a model for corporations is in its *willingness to embrace (the right kind of) failures*. I have highlighted at several points the wide range of outcomes of venture-backed projects, and the way in which the worthy failure is celebrated in Silicon Valley as an essential rite of passage for an aspiring entrepreneur. Far too often in corporations, on the other hand, any hint of a reversal is shunned.

To illustrate this point, one R&D executive who had started his career at Harvard Medical School described the process of ending a drug development program at his pharmaceutical company as more arduous than firing a tenured professor at Harvard. (Since his old employer had been unable to dismiss a professor notorious for appearing on talk shows discussing alien abductions, this was really saying something.) This aversion to failure manifests itself in corporate venturing as well: all too often, a team with a program that has operated for several years will proudly boast that it has not yet shut down any of the firms in its portfolio. While their corporate superiors may interpret this fact as a sign of success, given the nature of the entrepreneurial process, it is highly likely that the group is bankrolling some moribund firms—the "living dead," in venture parlance—whose funding should have been cut off years before.

Numerous models ensure the combination of short-term security (to ensure experimentation) and long-term rewards (to motivate hard work) that Gustavo Manso and others recommend. Certainly, the venture capital process embodies many of these features. Corporations would do well to model their internal venturing efforts on the approaches of independent groups, perhaps with modifications such as the S.R. One approach (discussed in chapter 6).

A more general problem faced by corporations is figuring out how to offer appropriate rewards for those within research laboratories. Incorrectly designed rewards can distort behavior. (Think back to the Xerox program described in chapter 1, where the presence of much greater rewards for spin-offs led some researchers to shift away from working on the firm's core products.) The ideal reward structure will both encourage progress in areas critically important to the firm and ensure that if multiple team members are necessary for working on a project, all contributors will be rewarded.

The key is to remove some of the stigma from the failure for corporate innovators—something easier said than done. Certainly, the R&D department of a firm where researchers are forgiven for every misstep and lack accountability will eventually be dominated by malingerers and incompetents. Rather, a critical step is to link rewards to performance, but over a time span that allows room for experiments, some of which are almost certain to be unsuccessful.

One crucial, though often neglected, point is that such tolerance for failure requires a rethinking not just of compensation schemes, but also of how projects are selected and funded. One of the reasons why failure is not an option in many corporate laboratories is that group leaders are loath to endanger continued funding for their projects. A question that would reward both further research by economic theorists and real-world exploration is how to induce "truth telling" when evaluating high-risk innovative projects. Several examples exist of ways to address this problem, from venture groups who employ a "devil's advocate" to make the case why a proposed investment should not be undertaken, to corporations who rely heavily on outside experts when making project funding decisions (e.g., the Glaxo scheme discussed in chapter 3). But far more could be done here.

One very fertile area for corporations to consider is the adoption of prize schemes within the company. As discussed in chapter 3, major corporations are increasingly adopting prize schemes, but they have mostly been used to encourage outsiders to address the firm's problems. Such external contests have the potential benefit of alerting the firm to approaches that may be different from those used internally, as well as generating awareness of and good publicity about the firm's innovative efforts.

But internal contests can also have some very real benefits:

- The contests can be designed flexibly, in a way that a firmwide compensation scheme cannot. For instance, these schemes can primarily reward long-term efforts that are of considerable strategic importance to the corporation.

- Prizes can be structured in a way that rewards teams. The Netflix Prize discussed in chapter 3 highlights the flexibility of these arrangements. While such joint efforts might be more challenging to adjudicate within a laboratory where there may be a large number of smaller contributions, the design of appropriate rewards should be possible.

- These rewards can satisfy the desire of researchers for both explicit monetary rewards and the intrinsic desire for recognition discussed earlier. One illustration of the dual role that prizes can play is in the study Liam Brunt, Tom Nicholas, and I conducted of the contests run by the Royal Agricultural Society of England between 1839 and 1939.[8] One of the most long-running innovation competitions worldwide, these events recognized a wide variety of innovations, from mechanized harvesters to new irrigation methods. We found that these prizes provided a powerful spur to innovation, even after controlling for the possibility that the Royal Society focused the competitions around particularly promising technological areas. Moreover, it was not simply the prospect of financial rewards that motivated the innovators: the impact of the rewards that offered simply recognition, such as the Society's Gold Medal, was also substantial.

Another important lesson relates to the importance of *investing in knowledge transfer*. Since the pioneering work of Wes Cohen and Dan Levinthal, students of innovation have understood that "absorptive capacity," or an ability to learn from others, is a critical success factor.[9] But far too often, though the individuals running a corporate venturing unit or a business development unit may learn valuable information, it does not get conveyed back to the operating units in a timely or usable manner. Because of this neglect, many of the most important potential gains from these efforts go to waste.

The barriers that prevent such transfers of knowledge are easy to understand. The corporate venturing and business development groups may be located far from the firm's operations. The former groups are likely to be dominated by young MBAs, while the operating managers may be far more seasoned engineers. The fact that corporate life—particularly in the current era of austerity—overwhelms managers with day-to-day responsibilities may dull the focus on building ties with others in the organization. Finally, many of the technologies being developed by the alliance or venture partners may be sufficiently protean that their applicability within the corporation is hard to discern.

Corporations have tried several approaches to overcome these barriers. One approach, employed by General Electric and others, has been to put an operating manager on the board of each portfolio firm. The rationale was that these individuals, by engaging with the young firms on a regular basis, would become conduits of knowledge back into the firm. Alas, such efforts have typically failed. A manager running a two-thousand-person refrigerator assembly plant is unlikely to have much time to worry about a ten-person start-up. The odds that the technology will address the assembly plant's specific problems are low, and the manager is unlikely to have the energy to hunt down who in the organization would be a better fit. Moreover, corporate middle managers typically have little expertise with the specific issues regarding entrepreneurial finance, strategy, and human resource management that start-up boards must regularly face.

A more successful approach has been to create a dedicated unit devoted to transferring the knowledge from the alliances and corporate investments back into the firm. Probably the best illustration along these lines is the Central Intelligence Agency's venture capital program, In-Q-Tel.[10] Founded in 1999 to acquire greater access to novel technologies on behalf of the U.S. intelligence community, the fund primarily made equity investments in cutting-edge technologies developed by young firms. Often, these firms had developed products for the private sector without considering their national security applications. For instance, one of the first companies In-Q-Tel funded was Las Vegas–based Systems Research Development (SRD), a producer of data analysis software for identifying "nonobvious" relationships. Jeff Jonas, SRD's founder and chief scientist, had written the nonobvious relationship analysis package for casino industry clients, allowing them to detect and thwart card counters and other cheaters. As Jonas explained, "It was very hard to be a West Coast company that's never done anything in Washington, with no visibility or awareness into sensitive federal organizations. You can't just show up from Vegas and say, 'Do you want to buy a watch?' "[11]

The knowledge transfer challenges were particularly challenging here. Not only could it be difficult for an outsider to identify which projects in the intelligence community would find relationship software useful, but substantial adaptation of the technologies was required in order to move from identifying MIT students at the Caesars Palace blackjack tables to locating al Qaeda members in Yemen. Moreover, communications between the start-up executives and the product developers within the agency were severely constrained by limits on communicating classified information.

To address this challenge, In-Q-Tel adopted a bipartite structure. The firm's Silicon Valley–based venture team consisted of approximately twelve employees, closely mirroring a traditional group where general partners and associates scouted deals, performed due diligence, prepared term sheets, and shepherded portfolio companies. At the same time, the company's twenty-employee technology team worked in Arlington, Virginia. The latter group focused their efforts on assessing new technologies, testing the appropriateness of portfolio firms' technologies for the agency, and interacting with intelligence officials. They also helped agency officials prepare demonstrations of target technologies and provided assessments of the government's needs to guide the investors. Unlike the venture team, which was dominated by former entrepreneurs and recent graduates, many of the technology team members were seasoned technology executives with experience working with the intelligence community.

The problems that In-Q-Tel faced may be extreme, but they underscore a fundamental and often-neglected principle: knowledge flows from start-ups to large organizations are far from automatic. If corporations really want to enjoy the benefits from a hybrid model, they cannot leave these spillovers to chance. Only by investing in people who can play such a translational role are the knowledge transfers likely to happen.

Underlying many of the arguments above is a final key point: *the need for organizational experimentation.* Far too often, the corporate world does a poor job of learning from the past: earlier initiatives, if unsuccessful, are forgotten about, and the key architects dismissed or exiled to the Kazakh subsidiary. The outcome is highly predictable: many firms seem destined to repeat the past, making the same mistakes in pursuing innovation again and again.

An illustration is the experience of General Electric, which repeatedly began corporate venturing efforts, only to abandon them after the venture teams left due to frustration over the disconnect between their broad responsibilities and modest compensation levels.[12] For instance, in 1981 pioneering venture investor Pedro Castillo and his team left GE Business Development Services to establish Fairfield Venture Management; in 1998 and 1999, the program's successor, GE Equity, lost eighteen investors, a number of whom went on to lead groups at leading firms. In each of these cases—and a number of other, less publicized ones—the cause of the departure was the same: the refusal of

Jack Welch (and Reginald Jones before him) to allow compensation of the venture team to be linked to the performance of the firms in which they invested. "We can't have people in separate rowboats," Welch insisted, arguing, "We don't want anybody in our company going to a meeting with a different interest from everybody else."[13] A longtime observer of corporate venturing, Steve Kahn of Advent International, had a different explanation: "GE is a perfect example of how corporations are not set up to establish the same incentives as a private firm, because those people would be the highest paid in the firm."[14] However desirable a uniform system of compensation, General Electric's desire to attract top-tier investors while not paying even close to market salaries led to repeated and highly disruptive turnover.

This inability to design and learn from experiments runs counter to what we are seeing today in the entrepreneurial world. The lean start-up has attracted a great deal of attention as a route to commercialize new ideas rapidly.[15] As articulated by advocates like Eric Ries, the defining characteristics of these ventures are not that they are necessarily inexpensive, but rather that they rely on frequent iteration and careful attention to market feedback. Frequently, these firms will set up structured experiments, where decisions about product design will be made not by expert opinion (e.g., the layout of a website by a design expert), but rather by the discovery of which configuration has the greatest customer utilization (which of several dozen designs produces the most click-throughs). The design of a social media site is very different from that of a research laboratory: in the former case, one often gets feedback in minutes, while in the latter, assessing performance may take years or even decades. But despite these differences, the unwillingness of many corporations to revisit the past with an eye to understanding what went right and went wrong is very troubling.

The basic message is very simple: despite the academic and real-world insights of recent years, many aspects of the innovation process remain poorly understood. Rather than sticking to one time-honored route of pursuing new ideas, exploring the impact of different organizational structures—whether internal skunk works or formal corporate venturing initiatives—is likely to be a recipe for success.

To be successful, much of the same discipline found in the lean start-up movement will be required. To the extent possible, these comparisons should rely on randomized designs: that is, they should compare the innovative output from one set of projects (which employ one approach) to another group employing an alternative structure. The subset of projects that receive each approach would be randomly chosen. A simple randomized design like this is the most scientifically reliable method, and it is often the most cost-effective method for estimating the impact of a new organizational structure, as it allows us to learn the maximum amount by affecting the smallest number of researchers. This is the reason that randomization is so commonly used for testing new drugs and educational approaches where important interventions are evaluated. There is no reason that if randomized approaches are used to evaluate the safety of innovative products and processes, they cannot be used to assess the ways that innovation itself is organized.

Final Thoughts

Despite the vast amounts written about innovation over the years, our understanding of its drivers remains surprisingly limited. While organizational economists have made strides in understanding what combinations of incentives and organization structure can encourage innovative breakthroughs, many of these insights have not yet received the attention they deserve in the real world.

Given this backdrop, it is worth reemphasizing the final point: the power of experiments. The last decade has seen a myriad of new approaches toward innovation in the world of start-ups, among open source projects, in contests, and the like. Some of these new organizational models will doubtless end up as miserable failures, discarded in the dustbin of history. But other novel structures are already having profound impacts. The corporate innovation model is changing as well, but more slowly. Embracing a spirit of rigorous trial and error concerning the ways in which innovation is pursued is likely to yield substantial benefits, both to the corporate experimenter and to society as a whole.

Notes

Chapter 1

1. For a detailed summary of Motorola's patent reward scheme, see Mike Hughlett, "Motorola Curbs Chase for Patents," *Chicago Tribune*, August 21, 2005, http://articles.chicagotribune.com/2005-08-21/business/0508210220_1_patent-strategy-patent-office-motorola, and many web accounts.

2. Kevin C. Tofel, "Droid Does . . . Not Help Motorola Offset Market Share Loss," May 17, 2010, http://gigaom.com/2010/05/17/droid-does-not-help-motorola-offset-market-share-loss/.

3. See, for instance, the discussions in Ali Farhoomand and Kavita Sethi, "Motorola in China: Failure of Success?" Case HKU440 (Hong Kong: University of Hong Kong, 2005); and Alan MacCormack and Kerry Herman, "The Rise and Fall of Iridium," Case 601040 (Boston: Harvard Business School, 2001).

4. Steve Lohr, "A Bull Market in Tech Patents," *New York Times*, August 17, 2011, B1.

5. This account is based on, among other sources, Peter Coles and Tom Eisenmann, "Skype," Case 806-165 (Boston: Harvard Business School, 2009); Daniel Roth, "Catch Us If You Can," *Fortune*, February 9, 2004, 64–74; and "Skype Technologies, S.A.," Case EC37 (Stanford, CA: Stanford University Graduate School of Business, 2006).

6. This account is based on Henry Chesbrough, "The Governance and Performance of Xerox's Technology Spin-Off Companies," *Research Policy* 32 (2003): 403–421; Brian Hunt and Josh Lerner, "Xerox Technology Ventures: March 1995," Case 295127 and Teaching Note 298152 (Boston: Harvard Business School, 1995); and public security filings and press accounts.

7. Douglas K. Smith, *Fumbling the Future: How Xerox Invented, Then Ignored, the First Personal Computer* (New York: Morrow, 1988).

8. Larry Armstrong, "Nurturing an Employee's Brainchild," *BusinessWeek*, October 23, 1993, 196.

9. In figuring these gains, I assumed that Xerox sold its stakes in firms that went public at the time of the initial public offering, rather than the substantially appreciated prices thereafter, and valued investments whose outcome could not be determined at cost, less a 25% discount for illiquidity.

10. A fund at the 75th percentile of performance would have had a return of 17.7%. Thomson Reuters, "VentureXpert Database," http://www.venturexpert.com. The benchmark figures reported are capital-weighted averages.

11. This paragraph is based on "Xerox New Enterprises," http://www.xerox.com/xne (accessed via http://waybackmachine.org); and Nick Turner, "Xerox Inventions Now Raised Instead of Adopted by Others," *Investors' Business Daily*, January 28, 1997, A6.

12. These are archived at http://www.youtube.com/watch?v=TZb0avfQme8.

13. Mary Meeker, "USA Inc.: A Basic Summary of America's Financial Statements," February 2011, http://www.kpcb.com/usainc.

14. United States Congress, Congressional Budget Office, *The Long-Term Budget Outlook,* June 2011, http://www.cbo.gov/ftpdocs/122xx/doc12212/06-21-Long-Term_Budget_Outlook.pdf.

15. Morris Abramowitz, "Resource and Output Trends in the United States Since 1870," *American Economic Review* 46 (1956): 5–23; and Robert M. Solow, "Technical Change and the Aggregate Production Function," *Review of Economics and Statistics* 39 (1957): 312–320.

Chapter 2

1. Francis Bacon, *New Atlantis,* in *Francis Bacon: A Selection of His Works,* ed. Sidney Warhaft (New York: Baker, Vourhis, 1965). For a detailed discussion of Bacon's work and the broader evolution of proprietary R&D, see Ernan McMullin, "Openness and Secrecy in Science: Some Notes on Early History," *Science, Technology and Human Values* 10 (1985): 14–23.

2. Bacon, *New Atlantis,* 456.

3. Although we don't have comprehensive information, the role of industry R&D was about the same magnitude before World War II: estimates by David Mowery and Nathan Rosenberg suggest that corporations undertook at least two-thirds of the R&D in the United States during the 1930s, and even more beforehand, in *Technology and the Pursuit of Economic Growth* (New York: Cambridge University Press, 1991). Technically, they compute (unlike the earlier data) who provided the funding for the R&D, not who undertook it. Because federal funding of corporate research was quite modest before World War II, however, these two shares should be quite similar.

4. Based on U.S. National Science Board, *Science and Engineering Indicators: 2010* (Washington, DC: Government Printing Office, 2010), Appendix table 4-12, and similar tables in earlier editions.

5. Based on corporate securities filings in Standard & Poor's, "Compustat North America," http://www.compustat.com, and Thomson Reuters, "Worldscope," http://thomsonreuters.com/products_services/financial/financial_products/a-z/worldscope_fundamentals/, and includes public firms only.

6. One caveat is that all these numbers should be taken with a substantial grain of salt. Several limitations motivate this caution:

- R&D activity is typically estimated from government surveys and securities filings. Not only do definitions and methodologies vary across countries, but different firms may give very different answers, despite having similar circumstances. And finally, even at the largest firms, the person who fills out the government survey is likely to be a lower-level functionary.
- Small firms often do not have dedicated R&D teams, so much innovation takes place alongside day-to-day operations. These activities are harder to capture, and may be missed when they compile R&D spending.

Because in many nations, the tax credit for R&D applies only to manufacturers, service firms may not bother to track these activities. For instance, two financial firms over the past two decades that were among the most aggressive in hiring PhDs and introducing new products were Goldman Sachs and Merrill Lynch (before the financial crisis). Yet these firms typically reported no R&D at all. For a discussion and illustration of these issues, see Lawrence D. Brown, Thomas J. Plewes, and Marisa A. Gerstein, eds., *Measuring Research and Development Expenditures in the U.S. Economy,* Unnumbered Committee on National Statistics Report (Washington, DC: National Academies Press, 2004).

7. Nathan Rosenberg, "The Impact of the Inventor: A Historical View," in *The Positive Sum Strategy: Harnessing Technology for Economic Growth*, eds. Ralph Landau and Nathan Rosenberg (Washington, DC: National Academies Press, 1986), 17–31.

8. Kenneth Arrow, "Economic Welfare and the Allocation of Resources for Invention," in *The Rate and Direction of Inventive Activity: Economic and Social Factors*, ed. Richard R. Nelson (Princeton, NJ: Princeton University Press for the National Bureau of Economic Research, 1962), 609–626.

9. Alfred Marshall, *Principles of Economics* (London: Macmillan, 1890).

10. Chris Liu, "A Spatial Ecology of Structural Holes: Scientists and Communication at a Biotechnology Firm," unpublished working paper, University of Toronto, 2011.

11. The next four paragraphs are drawn from Grace R. Cooper, *The Sewing Machine: Its Invention and Development* (Washington, DC: Smithsonian Institution Press, 1976); "Elias Howe, Libraries of Curious and Unusual Facts: Inventive Genius," Time-Life Books, http://www.history.rochester.edu/Scientific_American/mystery/howe.htm; Ryan L. Lampe and Petra Moser, "Do Patent Pools Encourage Innovation? Evidence from the 19th-Century Sewing Machine Industry," working paper 15061, National Bureau of Economic Research, 2009; and "Obituary: Elias Howe, Jr.," *New York Times*, October 5, 1867, 2.

12. Based on Naomi R. Lamoreaux and Kenneth L. Sokoloff, "The Decline of the Independent Inventor: A Schumpeterian Story?" working paper 11654, National Bureau of Economic Research, 2005, Figure 2. Inventors with a primary education are those who did not attend school after the age of twelve (or not at all).

13. P. J. Federico, "Operation of the Patent Act of 1790," *Journal of the Patent Office Society* 18 (1936): 237–251. In actuality, this process did not work well and was soon abandoned, but the primary failing was the fact that the cabinet officials who made up the commission had much bigger worries than adjudicating patent disputes.

14. Naomi R. Lamoreaux and Kenneth L. Sokoloff, "Inventors, Firms, and the Market for Technology in the Late Nineteenth and Early Twentieth Centuries," in *Learning by Doing in Markets, Firms, and Countries*, eds. Naomi R. Lamoreaux, Daniel M. G. Raff, and Peter Temi (Chicago: University of Chicago Press, 1999), 19–60.

15. For a discussion of the process of nineteenth-century patent reform, see Adam Jaffe and Josh Lerner, *Innovation and Its Discontents* (Princeton, NJ: Princeton University Press, 2004). For the complex history of the treatment of employee inventors, see C. L. Fisk, "Removing the 'Fuel of Interest' from the 'Fire of Genius': Law and the Employee-Inventor, 1830–1930," *University of Chicago Law Review* 65 (1998): 1128–1198.

16. Lamoreaux and Sokoloff, "The Decline of the Independent Inventor," Table 10.

17. See, for instance, Ernst Homburg, "The Emergence of Research Laboratories in the Dyestuffs Industry, 1870–1900," *British Journal for the History of Science* 25 (1992): 91–111; and Georg Meyer-Thurow, "The Industrialization of Invention: A Case Study from the German Chemical Industry," *Isis* 73 (1982): 363–381.

18. See David C. Mowery and Nathan Rosenberg, *Paths of Innovation: Technological Change in 20th-Century America* (New York: Cambridge University Press, 1998), Table 1. The original data is from periodic surveys by the National Research Council of industrial research employment.

19. David Hounshell and John Smith, *Science and Corporate Strategy: Du Pont R&D, 1902–1980* (Cambridge, UK: Cambridge University Press, 1988). This account is also drawn in part from William S. Dutton, *Du Pont: One Hundred and Forty Years* (New York: C. Scribner's, 1942).

20. Hounshell and Smith, *Science and Corporate Strategy*, 46.

21. Ibid., 50, 61.

22. See, for instance, the account of the challenges faced by Willis Whitney, the founder of the General Electric Research Laboratory, in George Wise, *Willis R. Whitney, General Electric, and the Origins of U.S. Industrial Research* (New York: Columbia University Press, 1985).

23. Hounshell and Smith, *Science and Corporate Strategy*, 223.

24. Ibid., 229.

25. The first three paragraphs in this section are based on David A. Hounshell, "The Evolution of Industrial Research in the United States," in *Engines of Innovation: U.S. Industrial Research at the End of an Era*, eds. Richard S. Rosenbloom and William J. Spencer (Boston: Harvard Business School Press, 1996), 13–85.

26. Hounshell, "Evolution," 50.

27. For discussions, see Giovanni Gavetti, Rebecca Henderson, and Simona Giorgi, "Kodak and the Digital Revolution (A)," Case 703503 (Boston: Harvard Business School, 2004); and Mary J. Benner and Mary Tripsas, "The Influence of Prior Industry Affiliation on Framing in Nascent Industries: The Evolution of Digital Cameras," *Strategic Management Journal*, forthcoming.

28. Joseph Bailey, "General Introduction," in *Administering Research and Development: The Behavior of Scientists and Engineers in Organizations*, eds. Charles D. Orth III, Joseph C. Bailey, and Francis W. Wolek (Homewood, IL: Richard D. Irwin, 1964), 5.

29. National Science Board, Committee on Industrial Support for R&D, *The Competitive Strength of U.S. Industrial Science and Technology: Strategic Issues*, Report NSG 92-138 (Washington, DC: National Science Foundation, 1992), iii.

30. Mike Jensen, "The Modern Industrial Revolution, Exit, and the Failure of Internal Control Systems," *Journal of Finance* 48 (1993): 831–880. These numbers are taken from the updated and corrected version posted at http://courses.essex.ac.uk/ac/ac928/jensenfailureinternal%20control.pdf. By new capital expenditures, I mean capital expenditures in excess of the depreciation reported in each year.

31. Bronwyn H. Hall, "The Stock Market's Valuation of R&D Investment During the 1980's," *American Economic Review Papers and Proceedings* 83 (May 1993): 259–264.

Chapter 3

1. This is drawn primarily from Stephen Baker, "Netflix Prize: Another Million at Stake," *BusinessWeek*, September 22, 2009, 6; James Bennett and Stan Lanning, "The Netflix Prize," in *Proceedings of KDD Cup and Workshop 2007* (San Jose, CA, August 12, 2007), 3–6; Steve Lohr, "A $1 Million Research Bargain for Netflix, and Maybe a Model for Others," *New York Times*, September 22, 2009, B1; and Eliot Van Buskirk, "How the Netflix Prize Was Won," *Wired*, September 22, 2009, http://www.wired.com/epicenter/2009/09/how-the-netflix-prize-was-won.

2. This account is based on Amy K. Karlson, Benjamin B. Bederson, and John SanGiovanni, "AppLens and LaunchTile: Two Designs for One-Handed Thumb Use on Small Devices," Technical Report HCIL-2004-37, 2004, http://hcil.cs.umd.edu/trs/2004-37/2004-37.html; Josh Lerner and Ann Leamon, "Microsoft IP Ventures," Case 810096 (Boston: Harvard Business School); John SanGiovanni and David C. Robbins, "Single-Handed Approach for Navigation of Application Tiles Using Panning and Zooming," U.S. Patent Application no. 20060190833, February 18, 2005, amended August 24, 2006; and various press accounts.

3. Chris Ziegler, "Microsoft Spinoff ZenZui's 'Zooming User Interface,' " engadget immobile, March 27, 2007, http://mobile.engadget.com/2007/03/27/microsoft-spinoff-zenzuis-zooming-user-interface/. For contemporaneous product demonstration videos, see http://www.cs.umd.edu/hcil/mobile/launchtile.mov and http://www.youtube.com/watch?v=r12eUXJNbl8&feature=related.

4. Lerner and Leamon, "Microsoft IP Ventures," p. 10.

5. These claims are based on a survey run by the Industrial Research Institute between 1992 and 1999. See Alden S. Bean, Jean Russo, Roger L. Whiteley, and assorted others, "Benchmarking Your R&D: Results from IRI/CIMS Annual R&D Survey," *Research Technology Management*, various years. The data from 1993 to 1998 are constructed to use a consistent set of firms; 1992 uses a somewhat different sample.

6. The quote is from John E. Rigdon and Joann S. Lubin, "Kodak Seeks Outsider to be Chairman, CEO," *Wall Street Journal*, August 9, 1993, A3. The R&D data is from Kodak's annual reports, compiled at U.S. Securities and Exchange Commission, "EDGAR Database," http://www.sec.gov/edgar.shtml.

7. This account of Lucent's history is drawn from Lisa Endlich, *Optical Illusions: Lucent and the Crash of Telecom* (New York: Simon & Schuster, 2004); Mary M. Frank and Jonathan M. Right, "Alcatel S.A. and Lucent Technologies," Case UV0730-PDF-ENG (Charlottesville, VA: Darden School of Business, 2007); and numerous press accounts.

8. Geoff Brumfield, "Bell Labs Bottoms Out: Institute Pulls Plug on Basic Research," *Nature*, August 20, 2008, 927.

9. Tom Randall, "Pfizer to Cut Research, Shut Plants in Savings Effort," *Bloomberg Business Week Online*, February 1, 2011, http://www.businessweek.com/news/2011-02-01/pfizer-to-cut-research-shut-plants-in-savings-effort.html.

10. Sarah Houlton, "Small Is Beautiful for GSK Drug Discovery," *Chemistry World*, July 25, 2008, http://www.rsc.org.ezp-prod1.hul.harvard.edu/chemistryworld/News/2008/July/25070805.asp; Toby Stuart and James Weber, "GSK's Acquisition of Sitris: Independence or Integration?" Case 809026 (Boston: Harvard Business School, 2009); and Jeanne Whalen, "To Innovate, Glaxo Brings Biotech Precepts In-House," *Wall Street Journal*, July 1, 2010, A5.

11. Stuart and Weber, "GSK's Acquisition of Sitris," 3.

12. Ibid., 5.

13. For a more extensive discussion of these strategies, see Josh Lerner and Mark Schankerman, *The Comingled Code: Open Source and Economic Development* (Cambridge, MA: MIT Press, 2010).

14. Rob Spiegel, "HP Sows Seeds for WebOS Ecosystem," *E-Commerce Times*, February 6, 2011, http://www.technewsworld.com/story/72576.html; and Hewlett-Packard, "HP to Contribute webOS to Open Source: HP to Enable Creativity of the Community to Accelerate the Next-Generation Web-Centric Platform," press release, December 9, 2011.

15. Jeremy Stein, "Internal Capital Markets and the Competition for Corporate Resources," *Journal of Finance* 52 (1997): 111–133.

16. The classic documentation of these patterns is in Ralph Katz and Thomas J. Allen, "Investigating the Not Invented Here (NIH) Syndrome: A Look at the Performance, Tenure, and Communication Patterns of 50 R & D Project Groups," *R&D Management* 12 (1982): 7–20.

17. For illustrations of these arguments, see Raghuram G. Rajan, Henri Servaes, and Luigi Zingales, "The Cost of Diversity: Diversification Discount and Inefficient

Investment," *Journal of Finance* 55 (2000): 35–80; and David S. Scharfstein and Jeremy C. Stein, "The Dark Side of Internal Capital Markets: Divisional Rent-Seeking and Inefficient Investment," *Journal of Finance* 55 (2000): 2537–2564.

18. Amit Seru, "Firm Boundaries Matter: Evidence from Conglomerates and R&D Activity," *Journal of Financial Economics*, forthcoming.

19. The most widely used measure for quality is patent citations. Patent applicants and the patent office functionaries who examine the applications must indicate which earlier awards are the most relevant. Patents that are more cited in subsequent patent documents are typically interpreted as having more impact or as being more important than less-cited patents.

20. Morten Sorensen, Per Stromberg, and Josh Lerner, "Private Equity and Long-Run Investment: The Case of Innovation," *Journal of Finance* 66 (2011): 445–477.

21. A few exceptions exist, such as Germany, where a government commission determines the proper reward for employee inventions. Not surprisingly, this approach is fraught with issues as well. For a discussion of the law of employee inventors, see Robert P. Merges, "The Law and Economics of Employee Inventions," *Harvard Journal of Law & Technology* 13 (Fall 1999): 1–54; for a discussion of the German case and its peculiarities, Dietmar Harhoff, "Institutionalized Incentives for Ingenuity—Patent Value and the German Employees' Inventions Act," *Research Policy* 36 (2007): 1143–1162.

22. This account is based on *DDB Technologies v. MLB Advanced Media, LP,* 465 F.Supp.2d 657 (W.D. Tex. 2006), 517 F. 3d 1284 (Fed. Cir. 2008), 676 F. Supp. 2d (W.D. Tex. 2009).

23. See David E. Brown, *Inventing Modern America: From the Microwave to the Mouse* (Cambridge, MA: MIT Press, 2003); John M. Osepchuk, "The History of the Microwave Oven: A Critical Review," *Microwave Symposium Digest, 2009* (June 2009): 1397–1400; and John M. Osepchuk, "A History of Microwave Heating Applications," *IEEE Transactions on Microwave Theory and Techniques*, MTT-32 (1984): 1200–1224.

24. This account is based on, among other sources, "Court Dismisses Inventor's Patent Claim but Will Consider Reward," *Japanese Times Online*, September 20, 2002, http://search.japantimes.co.jp/cgi-bin/nn20020920a2.html; and Ian Rowley and Hiroko Tashiro, "A Green Light for Inventors in Japan," *Bloomberg BusinessWeek.com*, July 11, 2005, http://www.businessweek.com/magazine/content/05_28/b3942411.htm.

25. Frederik Neumeyer, *The Employed Inventor in the United States: R&D Policies, Law, and Practice* (Cambridge, MA: MIT Press, 1971). The quote below is on page 101.

26. Ravinder K. Jain and Harry C. Triandis, *Management of Research and Development Organizations: Managing the Unmanageable* (New York: Wiley, 1990). The quotation is from page 164. Among the reasons they offer are the possibility of inequity in pay leading to reduced cooperation across researchers and the possibility of employees "gaming" the system to garner higher pay.

27. Todd Zenger and Sergio Lazzarini, "Compensating for Innovation: Do Small Firms Offer High-Powered Incentives That Lure Talent and Motivate Effort?" *Managerial and Decision Economics* 25 (2004): 329–345.

28. See the discussions in Bengt Holmstrom, "Agency Costs and Innovation," *Journal of Economic Behavior and Organization* 12 (1989): 305–327; and Bengt Holmstrom and Paul Milgrom, "Multitask Principal-Agent Analyses: Incentive Contracts, Asset Ownership, and Job Design," *Journal of Law, Economics, and Organization* 7 (1991): 24–52.

29. Leonard R. Sayles, *Individualism and Big Business* (New York: McGraw Hill, 1963), chapter 14, 127–152.

30. Josh Lerner and Julie Wulf, "Innovation and Incentives: Evidence from Corporate R&D," *Review of Economics and Statistics* 89 (2007): 634–644.

31. It should be noted that these shifts are not unique to corporate R&D managers and mirror similar shifts in senior management compensation more broadly.

32. ipPerformance Group, *2011 Inventor Reward and Recognition Programs Benchmark* (Naperville, IL: ip Performance Group, 2011).

33. Casey Ichniowski, Kathy Shaw, and Giovanna Prennushi, "The Effects of Human Resource Management Practices on Productivity: A Study of Steel Finishing Lines," *American Economic Review* 87 (1997): 291–313.

34. The golf study alluded to is Ronald G. Ehrenberg and Michael L. Bognanno, "Do Tournaments Have Incentive Effects?" *Journal of Political Economy* 98 (1990): 1307–1324.

35. Teresa Amabile, "Social Psychology of Creativity: A Consensual Assessment Technique," *Journal of Personality and Social Psychology* 43 (1982): 997–1013.

36. For the classic discussion of these problems, see Bengt Holmstrom, "Agency Costs and Innovation." The quote is from Samuel R. Phillips, "What Is an Inventor's Fair Share?" *Machine Design* 67 (October 26, 1995): 163.

37. Gustavo Manso, "Motivating Innovation," *Journal of Finance*, forthcoming.

38. Richard Stallman, "The GNU Operating System and the Free Software Movement," in *Open Sources: Voices from the Open Source Revolution*, eds. Chris DiBona, Sam Ockman, and Mark Stone (Sebastopol, CA: O'Reilly, 1999), 53–70. The quotation is from page 54.

39. Eric Raymond, "Homesteading the Noosphere: An Introductory Contradiction," 2000, http://catb.org/~esr/writings/homesteading/homesteading/.

40. For a fuller discussion, see Josh Lerner and Jean Tirole, "Some Simple Economics of Open Source," *Journal of Industrial Economics* 52 (2002): 197–234.

41. Deborah Claymon, "Jobs Bans Credits on Apple Products," *San Jose Mercury News*, December 2, 1999, 1C.

42. Jeffrey Roberts, Il-Horn Hann, and Sandra Slaughter, "Understanding the Motivations, Participation, and Performance of Open Source Software Developers: A Longitudinal Study of the Apache Projects," *Management Science* 52 (2006): 984–999.

43. These results provide some clues why contributions to open source projects continue to be robust, even as alternative ways for programmers to signal their ability have emerged. In recent years, websites like Top Coder have run large-scale competitions for programmers to solve particular software problems posed by firms and individuals, in exchange for recognition and cash rewards. (For an overview of how these contests work and affect programmers, see Kevin J. Boudreau, Nicola Lacetera, and Karim R. Lakhani, "Incentives and Problem Uncertainty in Innovation Contests: An Empirical Analysis," *Management Science* 57 (2011): 843–863.) These contests, however, primarily test programming ability, rather than the more valuable project management skills that are needed to assume a leadership position in an open source project.

44. Florian Ederer and Gustavo Manso, "Is Pay-for-Performance Detrimental to Innovation?" working paper 936643, Institute for Business and Economic Research, Department of Economics, University of California at Berkeley, 2009.

45. Gustavo Manso, Pierre Azoulay, and Joshua Graff Zivin. "Incentives and Creativity: Evidence from the Academic Life Sciences," *Rand Journal of Economics*, forthcoming.

46. Lerner and Wulf, "Innovation and Incentives: Evidence from Corporate R&D."

Chapter 4

1. This account is drawn from a wide variety of sources, including "Suniva, Inc.: Business Overview," *Bloomberg BusinessWeek.com*, http://investing.businessweek. com/research/stocks/private/snapshot.asp?privcapId=40786864; and Ucilia Wang, "Suniva Bags $94M After Ditching Fed Loan Guarantee Plan," April 14, 2011, http:// www.reuters.com/article/2011/04/14/idUS11646255520110414.

2. The definitive account of the early history of investment partnerships is Guido Astuti, *Origini e Svolgimento Storico della Commenda Fino al Secolo XIII* (Milan: S. Lattes & Co., 1933). The best accounts in English are Raymond De Roover, "The Organization of Trade," in *The Cambridge Economic History of Europe: Volume III— Economic Organization and Policies in the Middle Ages*, eds. M. M. Postan, E. E. Rich, and Edward Miller (Cambridge, UK: Cambridge University Press, 1963), chapter 2; and Robert S. Lopez and Irving W. Raymond, *Medieval Trade in the Mediterranean World: Illustrative Documents Translated with Introductions and Notes* (New York: Columbia University Press, 1955).

3. The discussion of ARD in this chapter is based on, among other sources, Spencer E. Ante, *Creative Capital: Georges Doriot and the Birth of Venture Capital* (Boston: Harvard Business School Press, 2008); David H. Hsu and Martin Kenney, "Organizing Venture Capital: The Rise and Demise of American Research & Development Corporation, 1946–1973," *Industrial and Corporate Change* 14 (2005): 579–616; and Patrick R. Liles, *Sustaining the Venture Capital Firm* (Cambridge, MA: Management Analysis Center, 1977).

4. Quoted in Liles, *Sustaining the Venture Capital Firm*, 39.

5. Ibid, 40. The original quotation is from Doriot's 1949 letter to shareholders.

6. Steven Kaplan and Per Stromberg, "Characteristics, Contracts, and Actions: Evidence from Venture Capitalist Analyses," *Journal of Finance* 59 (2004): 2177–2210.

7. Steven Kaplan and Per Stromberg, "Financial Contracting Theory Meets the Real World: An Empirical Analysis of Venture Capital Contracts," *Review of Economic Studies* 70 (2003): 281–315.

8. Paul Gompers, "Optimal Investment, Monitoring, and the Staging of Venture Capital," *Journal of Finance* 50 (1995): 1461–1489.

9. Josh Lerner, "Venture Capitalists and the Oversight of Private Firms," *Journal of Finance* 50 (1995): 301–318.

10. This account is especially based on Leslie Berlin, "Draper Gaither and Anderson: First Venture Capital Firm in Silicon Valley," in *Making the American Century: Studies in 20th Century Culture, Politics, and Economy*, ed. Bruce Schulman (New York: Oxford University Press, forthcoming); William H. Draper Jr., *The Startup Game: Inside the Partnership Between Venture Capitalists and Entrepreneurs* (New York: Palgrave Macmillan, 2011); and "Blue Ribbon Venture Capital," *BusinessWeek*, October 29, 1960, 65–69.

11. "General Doriot's Dream Factory," *Fortune* 76 (August 1967): 103–136. The quote is on page 104.

12. Berlin, "Draper Gaither and Anderson," 4.

13. Draper, *The Startup Game*, 26.

14. "An Analysis of Compensation in the U.S. Venture Capital Partnership," *Journal of Financial Economics* 51 (1999): 3–44; and "The Use of Covenants: An Analysis of Venture Partnership Agreements," *Journal of Law and Economics* 39 (1996): 463–498.

15. This account is based on numerous press accounts, but especially Peter Brooke and Daniel Penrice, *A Vision for Venture Capital: Realizing the Global Promise of*

Venture Capital and Private Equity (Boston: New Ventures Press, 2009); Paul A. Gompers and Jeffrey Anapolsky, "Advent Israel Venture Capital Program," Case 298072 (Boston: Harvard Business School, 1998); and Elizabeth B. Stein and Debora L. Spar, "The Advent of Venture Capital in Latin America," Case 797077 (Boston: Harvard Business School, 1998). It should be noted that Brooke had several contemporaries in developing global venture investing, most notably the nearly simultaneous fund launched by Ronald Cohen and Alan Patricof (which would lead to the creation of Apax Partners).

16. Preqin, "Fund Manager Profile," http://www.preqin.com.

17. For two illustrative analyses, see Leslie A. Jeng and Philippe C. Wells, "The Determinants of Venture Capital Funding: Evidence Across Countries," *Journal of Corporate Finance* 6 (2000): 241–289; and Marco Da Rin, Giovanna Nicodano, and Alessandro Sembenelli, "Public Policy and the Creation of Active Venture Capital Markets," *Journal of Public Economics* 90 (2006): 1699–1723.

18. Preqin, "Fund Manager Profile."

19. Wade Roush, "Mike Maples and Ann Miura-Ko on The Limits of Incubators, the Right Fund Size, and the True Meaning of 'Pivot,' " *Xconomy*, June 13, 2011, http://www.xconomy.com/san-francisco/2011/06/13/mike-maples-and-ann-miura-ko-on-the-limits-of-incubators-the-right-fund-size-and-the-true-meaning-of-pivot/4.

20. This account is based on, among other sources, Paul Graham, "Graham on Start-ups, Innovation, and Creativity," audio blog interview by Russ Roberts, August 3, 2009, http://files.libertyfund.org/econtalk/y2009/Grahaminnovation.mp3; Paul Graham, "Y Combinator Numbers," June 2011, http://ycombinator.com/nums.html; and Steven Levy, "Y Combinator as Boot Camp for Startups," *Wired*, June 2011, 74–76.

21. Graham, "Graham on Start-Ups."

22. Ibid.

23. Steven Kaplan and Antoinette Schoar, "Private Equity Returns: Persistence and Capital Flows," *Journal of Finance* 60 (1995): 1791–1823.

24. This analysis is taken from Josh Lerner, Ann Leamon, and Felda Hardymon, *Private Equity, Venture Capital, and the Financing of Entrepreneurship: The Power of Active Investing* (New York: Wiley, 2012), chapter 12.

25. These are the predicted coefficients from a regression analysis, and thus represent the central tendency in data. Any individual fund may do better or worse.

26. The figures in this and the next paragraphs are based on the author's analysis of Thomson Reuters, "VentureXpert," http://www.venturexpert.com; with supplemental information from Standard & Poor's, "Compustat North America," http://www.compustat.com.

27. Thomas Hellmann and Manju Puri, "The Interaction Between Product Market and Financing Strategy: The Role of Venture Capital," *Review of Financial Studies* 13 (2000): 959–984.

28. Samuel Kortum and Josh Lerner, "Assessing the Contribution of Venture Capital to Innovation," *Rand Journal of Economics* 31 (2000): 674–692. One concern about the Hellmann-Puri paper is that venture capitalists may simply be more adept at identifying industries where innovation is about to boom, and not cause the innovation themselves. To try to address these concerns about interpretation, we look back over the industry's history. In particular, a major discontinuity in the recent history of the venture capital industry was the U.S. Department of Labor's clarification of the Employee Retirement Income Security Act in the late 1970s, a policy shift that freed pensions to invest in venture capital. This shift led to a sharp increase in the funds committed to venture

capital. This type of external change should allow us to figure out what the impact of venture capital was, because the policy change by the midlevel bureaucrats buried deep within an obscure subagency of the U.S. Department of Labor is unlikely to be related to how many entrepreneurial opportunities there were to be funded.

Chapter 5

1. IHS Global Insight and National Venture Capital Association, *Venture Impact: The Economic Importance of Venture Capital-Backed Companies to the U.S. Economy,* 6th edition (Arlington, Virginia: NVCA, 2011).

2. This comparison is drawn from Robert L. Banks and Patrick R. Liles, "The Charles River Partnership," Case 375075 (Boston: Harvard Business School, 1975); "Charles River Ventures," http://www.crv.com/; and Thomson Reuters, "VentureXpert Database," http://www.venturexpert.com.

3. The venture capital investment data is compiled from Thomson Reuters, "VentureXpert Database." The R&D data is from the U.S. National Science Foundation, *Science and Engineering Indicators 2010,* Tables 4-12 and 4-13, http://www.nsf.gov/statistics/seind10/. The value-added activity is from the Bureau of Economic Analysis, U.S. Department of Commerce, "Annual Revision of the National Income and Product Accounts," Survey of Current Business, August 2010, http://www.bea.gov/scb/pdf/2010/08%20August/0810_nipa-revision.pdf.

4. This account is based on, among other sources, Yuliya Chernova, "Kleiner Perkins Hopes to Turn a Black Eye into Black Gold," *Venture Capital Dispatch* (blog), *Wall Street Journal,* February 15, 2011, http://blogs.wsj.com/venturecapital/2011/02/15/kleiner-perkins-hopes-to-turn-a-black-eye-into-black-gold/; Russ Garland, "Terralliance Struggles While Looking for Oil," *Venture Capital Dispatch* (blog), *Wall Street Journal,* April 6, 2009, http://blogs.wsj.com/venturecapital/2009/04/06/searching-for-oil-proves-costly-for-terralliance/; Adam Lashinsky, "There Will Be Oil," *Fortune,* April 12, 2010, 84–94; and Adam Lashinsky, "Kleiner Bets the Farm: The Legendary Venture Firm Is Going Green—and Leaving Internet Deals to the Competition," *Fortune,* July 24, 2008, 96–104.

5. According to VentureXpert, Kleiner's sole earlier foray into the sector was an investment with half a dozen peers in a Texas oil-field services business in the 1970s, where the total amount invested was a mere $3 million.

6. Lashinsky, "There Will Be Oil," 84.

7. Sand Hill Econometrics, "The Sand Hill Index: All Industries Combined," unpublished tabulation, 2011. I thank Susan Woodward for access to this information.

8. These are based on Thomson Reuters, "VentureXpert Database."

9. Miguel Helft, "A Kink in Venture Capital's Gold Chain," *New York Times,* October 7, 2006, C1.

10. Renee Deger, "Disbursements on the Rise: Sharp Gains Raise Yellow Flag to Some Venture Capitalists," *Venture Capital Journal* 33 (December 1993): 29.

11. "Special Report—Capital Transfusion Renewal," *Venture Capital Journal* 20 (July 1980): 6–8 (quotation is from pages 6 and 7).

12. The return numbers quoted are capital-weighted averages and are from Thomson Reuters, "VentureXpert Database."

13. Based on Paul Gompers and Josh Lerner, "Money Chasing Deals? The Impact of Fund Inflows on Private Equity Valuations," *Journal of Financial Economics* 55 (2000): 281–325.

14. Based on Steven N. Kaplan and Antoinette Schoar. "Private Equity Performance: Returns, Persistence and Capital Flows," *Journal of Finance* 60 (2005): 1791–1823.

15. For a discussion of these patterns, see Josh Lerner, Ann Leamon, and Felda Hardymon, *Private Equity, Venture Capital, and the Financing of Entrepreneurship: The Power of Active Investing* (New York: Wiley, 2011); and Josh Lerner, Antoinette Schoar, and Wan Wongsunwai, "Smart Institutions, Foolish Choices: The Limited Partner Performance Puzzle," *Journal of Finance* 62 (2007): 731–764.

16. For detailed documentation of the compensation of venture capitalists, see Andrew Metrick and Ayako Yasuda, "The Economics of Private Equity Funds," *Review of Financial Studies* 23 (2010): 2303–2341.

17. This account is based on numerous press accounts, including Josh Sandberg, "Pegasus Biologics Is Insolvent; Closing Their Doors," *Ortho Spine News*, May 16, 2009, http://www.orthospinenews.com/pegasus-biologics-is-insolvent-closing-their-doors-2; Vita Reed, "Minnesota Company Picks Up Pieces of Failed Startup," *Orange County Business Journal*, June 13, 2009, 4; and "Synovis Orthopedic and Woundcare Past and Present, Formerly Pegasus Biologics," 2011, http://synovisorthowound.com/Past-and-Present.asp.

18. Rick Townsend, "Propagation of Financial Shocks: The Case of Venture Capital," unpublished working paper, Dartmouth College, Hanover, NH, 2011.

19. This account is based on Josh Lerner, "An Empirical Analysis of a Technology Race," *Rand Journal of Economics* 28 (1997): 228–247; and Josh Lerner, "Pricing and Financial Resources: An Analysis of the Disk Drive Industry, 1980–88," *Review of Economics and Statistics* 77 (1995): 587–598.

20. This data is from National Venture Capital Association, *2011 NVCA Yearbook*, 2011, http://www.nvca.org/index.php?option=com_docman&task=doc_download&gid=710&Itemid=93. Data is originally from Thomson Reuters.

21. This account is based on, among other sources, Tara Lachapelle and Dina Bass, "Skype Gets 40% Markup as Microsoft Surprised Owners," *Bloomberg.com News*, May 11, 2011, http://www.bloomberg.com/news/2011-05-11/skype-gets-40-markup-as-microsoft-surprised-owners-real-m-a.html; and Kevin Maney, "Skype: The Inside Story of the Boffo $8.5 Billion Deal," *Fortune*, July 25, 2011, 125–128.

22. Jeremy Stein, "Takeover Threats and Managerial Myopia," *Journal of Political Economy* 96 (1988): 61–80.

23. This history is based primarily on Richard Coopey and Donald Clarke, *3i: Fifty Years Investing in Industry* (Oxford, UK: Oxford University Press, 1995); Felda Hardymon, Ann Leamon, and Josh Lerner, "3i Group PLC," Case 803020 (Boston: Harvard Business School, 2003); Henry Rinjen, "Global Pursuit," *Upside*, October 2000, 90–94; and "3i Drops Early Stage," *European Venture Capital and Private Equity Journal* 150 (March 2008): 15.

24. Dan Gledhill, "3i Has the Balance to Surf the Net's New Wave," *The Independent*, February 20, 2000, B7.

25. This is based on Thomson Reuters, "Datastream Database," http://online.thomsonreuters.com/datastream/.

26. Benjamin Wootliff, "Tech Downturn Hits 3i Group," *The Telegraph*, May 18, 2001, http://www.telegraph.co.uk/finance/4490768/Tech-downturn-hits-3i-Group.html.

27. Paul A. Gompers, "Grandstanding in the Venture Capital Industry," *Journal of Financial Economics* 42 (1996): 133–156; and Peggy M. Lee and Sunil Wahal, "Grandstanding, Certification and the Underpricing of Venture Capital Backed IPOs," *Journal of Financial Economics* 73 (2004): 375–407.

28. Wikipedia, s.v. "dot com party," http://en.wikipedia.org/wiki/Dot_com_party.

29. Joshua S. Gans and Scott Stern, "The Product Market and the Market for 'Ideas': Commercialization Strategies for Technology Entrepreneurs," *Research Policy* 32 (2003): 333–350.

30. Noam Wasserman, "The Founder's Dilemma," *Harvard Business Review*, February 2008, 103–109.

31. Ilan Guedj and David Scharfstein, "Organizational Scope and Investment: Evidence from the Drug Development Strategies and Performance of Biopharmaceutical Firms," working paper 10933, National Bureau of Economic Research, 2004.

32. Robert Hall and Susan Woodward, "The Burden of the Nondiversifiable Risk of Entrepreneurship," *American Economic Review* 100 (2010): 1163–1194. Making these calculations even grimmer is the fact that the two-decade period they examined included the "golden years" of the 1990s. A similar calculation over the last decade alone would surely produce even lower returns.

Chapter 6

1. Alexander Haislip, "Ackerman's Advice for Corporate Venture Capitalists," *Forbes Blogs*, February 28, 2011, http://blogs.forbes.com/alexanderhaislip/2011/02/28/ackermans-advice-for-corporate-venture-capitalists/.

2. These are based on Robert E. Gee, "Finding and Commercializing New Businesses," *Research/Technology Management* 37 (January/February 1994): 49–56; Paul Gompers and Josh Lerner, "The Determinants of Corporate Venture Capital Success: Organizational Structure, Incentives, and Complementarities," in *Concentrated Corporate Ownership*, ed. Randall Morck (Chicago: University of Chicago Press for the National Bureau of Economic Research, 2000), 17–50; and National Venture Capital Association, "Corporate Venture Capital Statistics," July 2011, http://www.nvca.org/index.php?option=com_docman&task=cat_view&gid=99&Itemid=317.

3. This description is based on Brian Gormley, "Eli Lilly's VC Arm Spins Out with $200M Fund," *VentureWire*, August 3, 2009, https://www.fis.dowjones.com/WebBlogs.aspx?aid=DJFVW00020090803e58300001&ProductIDFromApplication=&r=wsjblog&s=djfvw; Ron Laufer, David Lane, and Richard G. Hamermesh, "Corporate Venture Capital at Eli Lilly," Case 806092 (Boston: Harvard Business School, 2007); and other press accounts.

4. The account is based on G. Felda Hardymon and Ann Leamon, "Intel 64 Fund," Case 800351 (Boston: Harvard Business School, 2000); Barbara J. Mack, Adriana Boden, Lee Rand, and David Yoffie, "Intel Capital," Case 705408 (Boston: Harvard Business School, 2007); and numerous press accounts.

5. Intel Capital, "Intel Capital Invests $24.5 Million Across the Computer Continuum," May 23, 2011, http://www.intc.com/releases.cfm?ReleasesType=Intel+Capital.

6. This account is based on David H. Knights, "Analog Devices Enterprises/Bipolar Integrated Technology," Case 286117 (Boston: Harvard Business School, 1985); Rosabeth Moss Kanter, Jeffery North, Ann Piaget Bernstein, and Alistair Williamson, "Engines of Progress: Designing and Running Entrepreneurial Vehicles in Established Companies," *Journal of Business Venturing* 5 (1990): 415–430; Bruce G. Posner, "Mutual Benefits," *Inc.*, June 1984, 83–92; and various press accounts and securities filings.

7. Mack et al., "Intel Capital."

8. This paragraph is based on numerous press accounts, including Timothy Hay, "Something Ventured: VCs Watching Apple's App Store Closely," Dow Jones News

Service, July 16, 2008, www.dowjones.com/factiva; and Miguel Helft, "A Fund to Invest in Social Start-Ups," *New York Times*, October 25, 2010, B4.

9. Venture Economics, "Exxon," Corporate Venture Capital Study, unpublished manuscript, 1986; and Hollister B. Sykes, "Lessons from a New Ventures Program," *Harvard Business Review*, May–June 1986, 63–74.

10. Based on National Venture Capital Association, "Corporate Venture Capital Statistics."

11. Laufer et al., "Corporate Venture Capital at Eli Lilly."

12. See, for instance, David G. Barry, "S.R. One's Gavin Urges Newcomers to Stay in Game," *Corporate Venturing Report* 2 (January 2001): 1, 22; and "Portfolio Profiles: S.R. One," *Venture Capital Journal* 36 (October 1996): 39–41.

13. Gary Dushnitsky and Zur Shapira, "Entrepreneurial Finance Meets Organizational Reality: Comparing Investment Practices and Performance of Corporate and Independent Venture Capitalists," *Strategic Management Journal* 31 (2010): 990–1017.

14. Daniel Kang and Vik Nanda, "Complements or Substitutes? Technological and Financial Returns Created by Corporate Venture Capital Investments," unpublished working paper, Georgia Institute of Technology, July 23, 2011, http://papers.ssrn.com/sol3/papers.cfm?abstract_id=1893710.

15. Gompers and Lerner, "The Determinants of Corporate Venture Capital Success."

16. Thomas Chemmanur, Elena Loutskina, and Xuan Tian, "Corporate Venture Capital, Value Creation, and Innovation," unpublished working paper, Boston College, University of Virginia, and Indiana University, May 1, 2011, http://papers.ssrn.com/sol3/papers.cfm?abstract_id=1364213.

17. Steven Klepper, "The Capabilities of New Firms and the Evolution of the US Automobile Industry," *Industrial and Corporate Change* 11 (2002): 645–666; and Aaron K. Chatterji, "Spawned with a Silver Spoon? Entrepreneurial Performance and Innovation in the Medical Device Industry," *Strategic Management Journal* 30 (2009): 185–206.

18. Paul Gompers, David Scharfstein, and Josh Lerner, "Entrepreneurial Spawning: Public Corporations and the Formation of New Ventures, 1986–1999," *Journal of Finance* 60 (April 2005): 577–614.

19. See, for instance, the evidence in Suzanne E. Majewski, "Causes and Consequences of Strategic Alliance Formation: The Case of Biotechnology" (PhD diss., Department of Economics, University of California at Berkeley, 1998).

20. Philippe Aghion and Jean Tirole, "On the Management of Innovation," *Quarterly Journal of Economics* 109 (1994): 1185–1207. These ideas are tested and corroborated in Josh Lerner, Hilary Shane, and Alexander Tsai, "Do Equity Financing Cycles Matter? Evidence from Biotechnology Alliances," *Journal of Financial Economics* 67 (2003): 411–446.

21. Reinhard Angelmar and Yves Doz, "Advanced Drug Delivery Systems: ALZA and Ciba-Geigy," unnumbered INSEAD case study series, 1987–1989.

22. Ben Gomes-Casseres, Adam Jaffe, and John Hagedoorn, "Do Alliances Promote Knowledge Flows?" *Journal of Financial Economics* 80 (2006): 5–33.

Chapter 7

1. This account is based on, among other sources, Julie Creswell, "So Small a Town, So Many Patent Suits," *New York Times*, September 24, 2006, C1; Yan Leychkis, "Of Fire Ants and Claim Construction: An Empirical Study of the Meteoric Rise of the Eastern District of Texas as a Preeminent Forum for Patent Litigation," *Yale Journal*

of Law and Technology 9 (Spring–Fall 2006): 193–233; Timothy C. Meece and V. Bryan Medlock Jr., "Is TS Tech the Death Knell for Patent Litigation in the Eastern District of Texas?" unpublished working paper, Banner & Witcoff, Ltd. and Sidley Austin LLP, 2011, http://www.bannerwitcoff.com/_docs/library/articles/VBM%20Paper%20re_%20 Patent%20Litigation%20in%20E%20D%20%20Tex%20.pdf; and Sam Williams, "A Haven for Patent Pirates," *Technology Review Online*, 2006, http://www.technologyreview.com/article/16546/.

2. LegalMetric, "Top Five Patent Report," 2009, http://www.legalmetric.com/top5reports/.

3. Kimberly Moore, "Xenophobia in American Courts," *Northwestern University Law Review* 97 (2002–2003): 1497–1550.

4. Meece and Medlock, "Is TS Tech the Death Knell?"

5. Alton Pryor, *The Timeless Quotations of President Ronald Reagan* (Rosewell, CA: Stagecoach Publishing, 2005), 22.

6. Josh Lerner and Antoinette Schoar, "Does Legal Enforcement Affect Financial Transactions? The Contractual Channel in Private Equity," *Quarterly Journal of Economics* 120 (2005): 223–246.

7. For an extended discussion of these changes, see Adam B. Jaffe and Josh Lerner, *Innovation and Its Discontents: How Our Broken Patent System Is Endangering Innovation and Progress, and What to Do About It* (Princeton, NJ: Princeton University Press, 2004).

8. Based on Administrative Office of the United States Courts, *Director's Annual Report* (Washington, DC: Administrative Office, various years).

9. 547 U.S. 388 (2006).

10. Harold Wegner, "Leahy Smith America Invents Act (con'd): Fee Diversion May be Inevitable," e-mail correspondence, August 11, 2011.

11. David Drummond, "When Patents Attack Android," *The Official Google Blog*, August 4, 2011, http://googleblog.blogspot.com/2011/08/when-patents-attack-android .html.

12. The lessons from this case attracted extensive discussion in the business press and the academic literature. Two examples are Eric B. Chen, "Applying the Lessons of Re-Examination to Strengthen Patent Post-Grant Opposition," *Computer Law Review and Technology Journal* 10 (Winter 2006): 193–206; and Kenneth Terrell, "Patently Problematic: BlackBerry Case Spotlights Flaws in Software Review Process," *U.S. News and World Report*, February 19, 2006, 35–36.

13. Dabney Carr, "Senate Passes Patent Reform Without Litigation-Related Features," *Virginia IP Law Blog*, March 14, 2001, http://www.virginiaiplaw.com/2011/03/ articles/patent-litigation/senate-passes-patent-reform-without-litigationrelated- provisions/.

14. Josh Lerner, Lee Branstetter, and Takeshi Nakabayashi, "New Business Investment Company: October 1997," Case 299025 (Boston: Harvard Business School, 1999), provides a detailed discussion of policies in Japan during the 1990s. See also Curtis J. Milhaupt, "The Market for Innovation in the United States and Japan: Venture Capital and the Comparative Corporate Governance Debate," *Northwestern University Law Review* 91 (1996–1997): 865–898.

15. "FailCon 2011," http://www.failcon2010.com/.

16. Raymond Westbrook, "Slave and Master in Ancient Near Eastern Law," *Chicago-Kent Law Review* 70 (1995): 1631–1676. Today, the number of people in debt

bondage is estimated to be about 20 million. (Siddharth Kara, "A $110 Loan, Then Twenty Years of Debt Bondage," CNN, June 2, 2011, http://thecnnfreedomproject.blogs. cnn.com/2011/06/02/a-110-loan-then-20-years-of-debt-bondage/.)

17. See, for instance, "The New Debtors' Prisons," *New York Times*, April 5, 2009, A24.

18. Viral Acharya and Krishnamurthy Subramanian, "Bankruptcy Codes and Innovation," *Review of Financial Studies* 22 (2009): 4949–4988.

19. Augustin Landier, "Entrepreneurship and the Stigma of Failure," unpublished working paper, New York University, 2006.

20. These statistics are taken from U.S. National Science Foundation, "Graduate Students and Postdoctorates in Science and Engineering: Fall 2008, Detailed Statistical Tables," NSF Report 11-311, July 2011, http://www.nsf.gov/statistics/nsf11311/pdf/nsf11311.pdf, Tables 54, 55, and 70.

21. Vivek Wadhwa, AnnaLee Saxenian, Ben Rissing, and Gary Gereffi, "America's New Immigrant Entrepreneurs: Part I," Science, Technology & Innovation Paper 23, Duke University, January 2007, http://ssrn.com/abstract=990152.

22. Based on "Start-Up Chile," 2011, http://www.startupchile.org/; Nick Leiber, "Chile Aims to Attract Foreign Startups," *BusinessWeek*, October 25, 2011, 64; and Vivek Wadhwa, "Chile Wants Your Poor, Your Huddled Masses, Your Tech Entrepreneurs," *TechCrunch*, October 10, 2009, http://techcrunch.com/2009/10/10/chile-wants-your-poor-your-huddled-masses-your-tech-entrepreneurs/.

23. Reena Jana, "South America's New IT Hub," *BusinessWeek*, October 25, 2009, 73.

24. Ron Gilson, "The Legal Infrastructure of High Technology Industrial Districts: Silicon Valley, Route 128, and Covenants Not to Compete," *New York University Law Review* 74 (1999): 575–629. The quote is on page 579.

25. This work is reviewed in Matt Marx, "Non-compete Agreements: Barriers to Entry . . . and Exit?" *Innovation Policy and the Economy* 12 (2012): forthcoming.

26. Adam Jaffe and Josh Lerner, "Reinventing Public R&D: Patent Policy and the Commercialization of National Laboratory Technologies," *Rand Journal of Economics* 32 (2001): 167–198.

27. Jim Poterba, "Venture Capital and Capital Gains Taxation," *Tax Policy and the Economy* 3 (1989): 47–67.

28. Organisation for Economic Co-operation and Development, "Venture Capital Policy Review: United Kingdom," STI Working Paper 2003/1, Paris, OECD, 2003.

29. Based on Thomson Reuters, "SDC Platinum Database," http://thomsonreuters .com/products_services/financial/financial_products/a-z/sdc/.

30. Peter Iliev, "The Effect of SOX Section 404: Costs, Earnings Quality, and Stock Prices," *Journal of Finance* 65 (2010): 1153–1196.

31. This account is drawn from numerous reports, including Susanne Dirks and Mary Keeling, "Russia's Productivity Imperative: Leveraging Technology and Information to Drive Growth," Executive Report: IBM Institute for Business Value, September 2009, http://www.ibm.com/smarterplanet/global/files/us_en_us_gov ernment_gbe03244usen.pdf; Dmitry Medvedev, "Go Russia!" Russian Presidential Executive Office, September 10, 2009, http://eng.news.kremlin.ru/news/298; and Organisation for Economic Co-operation and Development, *OECD Reviews of Innovation Policy: Russian Federation* (Paris: OECD, 2011).

32. Medvedev, "Go Russia!"

33. Vannevar Bush, *Science, The Endless Frontier* (Washington, DC: U.S. Government Printing Office, 1945).

34. For a review, see European Commission, *Lisbon Strategy Evaluation Document*, Commission Staff Working Document SEC (2010) 114, February 2010, http://ec.europa .eu/europe2020/pdf/lisbon_strategy_evaluation_en.pdf. In actuality, Europe's progress in boosting R&D spending over the course of the decade was minimal.

35. For an overview of the program, see Gil Avnimelech, Martin Kenney, and Morris Teubal, "Building Venture Capital Industries: Understanding the U.S. and Israeli Experiences," BRIE Working Paper 160, 2004, http://brie.berkeley.edu/publications/wp160 .pdf; Organisation for Economic Co-operation and Development, *Venture Capital Policy Review: Israel*, STI Working Paper 2003/3, Paris, OECD, 2003; and Manuel Trajtenberg, "Government Support for Commercial R&D: Lessons from the Israeli Experience," *Innovation Policy and the Economy* 2 (2002): 79–134.

36. PricewaterhouseCoopers, "MoneyTree Report," https://www.pwcmoneytree .com/MTPublic/ns/index.jspcewater. Their compilations, however, do not include regions in China and India.

37. Yigdal Erlich, "The Yozma Group—Policy and Success Factors," http://www .insme.org/documenti/Yozma_presentation.pdf.

38. See the discussion and cases in Paul A. Gompers and Josh Lerner, *Capital Formation and Investment in Venture Markets: A Report to the NBER and the Advanced Technology Program,* Report GCR-99-784 (Washington, DC: Advanced Technology Program, National Institutes of Standards and Technology, U.S. Department of Commerce, 1999).

39. See, for instance, David M. Gold, "Cleantech Stimulus Not Very Stimulating," *GreenGold Blog*, September 29, 2009, http://www.greengoldblog.com/2009/09/clean tech-stimulus-not-very-stimulating.html; Scott Kirsner, "Does Lobbying Always Pay?" *Innovation Economy Blog*, August 6, 2009, http://www.boston.com/business/tech nology/innoeco/2009/08/does_lobbying_always_pay.html; Tim Mullaney, "Lobbyists Are First Winners in Obama's Clean-Technology Push," Bloomberg.com News, March 25, 2009, http://www.bloomberg.com/apps/news?pid=20601109&sid=aNH.vsK2D .lQ&refer=home; and Sean Sposito, "A123 Gets $249m in Stimulus Funding," *Boston Globe*, August 6, 2009, B5.

40. PriceWaterhouseCoopers, "MoneyTree Report."

41. See the discussions, for instance, in F. Thomson Leighton, "Presentation of the Draft Report on Cyber Security. President's Information Technology Advisory Committee," January 12, 2005, http://www.itrd.gov/pitac/meetings/2005/20050112/20050112_ leighton.pdf; John Markoff, "Pentagon Redirects Its Research Dollars," *New York Times*, April 2, 2005, C1; and National Research Council, *Assessment of Department of Defense Basic Research* (Washington, DC: National Academies Press, 2005).

42. "A Time for Choosing: A Speech to the 1964 Republican National Convention," October 27, 1964, http://www.reagan.utexas.edu/archives/reference/timechoosing .html. Reagan apparently borrowed the idea from former Senator James F. Byrnes.

43. See the history related in Bronwyn Hall and John Van Reenen, "How Effective Are Fiscal Incentives for R&D? A Review of the Evidence," *Research Policy* 29 (2000): 449–469. Among the most important of the critiques were Robert Eisner, Steven H. Albert, and Martin A. Sullivan, "The New Incremental Tax Credit for R&D: Incentive or Disincentive," *National Tax Journal* 37 (1986): 171–183.

44. More technically, the new base was the product of the firm's R&D-to-sales ratio between 1984 and 1988, and the revenues in the most recent years. Special formulas involving a fixed R&D-to-sales ratio were introduced for start-ups.

Chapter 8

1. This section is based in part on a presentation that Antoinette Schoar and I did for the Ewing Marion Kauffman Foundation conference titled "The Future of Venture Finance" in January 2010. Ed Colloton and Felda Hardymon contributed several helpful ideas to the presentation.

2. For evidence about the frequency of these patterns over time, see SCM Strategic Capital Management AG, *Annual Review of Private Equity Terms and Conditions*, Zurich, SCM, various years.

3. More technically, the net present values are reported, with an appropriate discount rate for each set of cash flows.

4. Of course, these sums were received over time, but this is present value of these payments at the time of the closing. Some of these payments would go to offset the expenses of running the fund.

5. Frank Angella, Felda Hardymon, Ann Leamon, and Josh Lerner, "Grove Street Advisors," Case 804050 (Boston: Harvard Business School, 2004).

6. This account is based on Jennifer Harris and Kevin Ley, "General Atlantic: A Global Firm with a Global Fund," *PEI Manager* 5 (December 2008): 19; Toby Lewis, "General Atlantic Deals Put It in the Big League," *Financial News*, June 21, 2010, http://www.efinancialnews.com/story/2010-06-21/general-atlantic-deals-put-it-in-the-big-league; and James Politi, "Private Equity Outsider That's in Step with Wall Street," *Financial Times*, July 18, 2006, 23.

7. Lewis, "General Atlantic Deals."

8. Liam Brunt, Tom Nicholas, and Josh Lerner, "Inducement Prizes and Innovation," working paper DP6917, Centre for Economic Policy Research, 2009.

9. Wes Cohen and Dan Levinthal, "Absorptive Capacity: A New Perspective on Learning and Innovation," *Administrative Science Quarterly* 35 (1990): 128–152. For a perspective on how the understanding of this phenomenon has evolved in recent years, see Kwanghui Lim, "The Many Faces of Absorptive Capacity: Spillovers of Copper Interconnect Technology for Semiconductor Chips," *Industrial and Corporate Change* 18 (2010): 1249–1284.

10. This account is based on Kevin Book, Felda Hardymon, Ann Leamon, and Josh Lerner, "In-Q-Tel," Case 804146 (Boston: Harvard Business School, 2005); Business Executives for National Security, *Accelerating the Acquisition and Implementation of New Technologies for Intelligence: The Report of the Independent Panel on the Central Intelligence Agency In-Q-Tel Venture* (Washington, DC: BNES, 2001); and Steven Levy, "Geek War on Terror," *Newsweek*, March 22, 2004, E6-E12.

11. Book et al., "In-Q-Tel," 11.

12. This account is drawn from, among other sources, Stephen Barr, "Bright Lights, Big Paychecks," *CFO*, March 1, 2000, 115–117; G. Felda Hardymon, Mark J. DeNino, and Malcolm S. Salter, "When Corporate Venture Capital Doesn't Work," *Harvard Business Review*, May/June 1983, 114–120; and Venture Economics, "General Electric," *Corporate Venture Capital Study*, unpublished manuscript, 1986.

13. Andrew Hill, "GE Chief Rules on Rewards," *Financial Times*, November 5, 1999, 25.

14. George Moriarity, "The New Smart Money: Once Considered Dumb, Corporate Venture Capital Is Now Sought by Entrepreneurs," *Investment Dealers Digest*, November 22, 1999, 16–21 (the quote is on page 21).

15. Steve Lohr, "The Rise of the Fleet-Footed Start-Up," *New York Times*, April 25, 2010, B5.

Index

About the Author

Josh Lerner is the Jacob H. Schiff Professor of Investment Banking at Harvard Business School, with a joint appointment in the Entrepreneurial Management and Finance units. His research focuses on innovation, venture capital, and private equity. He codirects the Productivity, Innovation, and Entrepreneurship Program at the National Bureau of Economic Research and also serves as coeditor of the bureau's annual publication, *Innovation Policy and the Economy*. Lerner founded and runs the Private Capital Research Institute, a nonprofit devoted to encouraging data access to and research about venture capital and private equity. In the 1993–1994 academic year, he introduced an elective course for second-year MBAs on private equity finance, which is consistently one of the largest elective courses at Harvard Business School. Lerner also teaches a doctoral course on entrepreneurship at HBS and runs the school's Owners/President Management Program. In 2010 he was the winner of the Swedish government's prestigious Global Award for Entrepreneurship Research.